Advances in Intelligent Systems and Computing

Volume 702

Series editor

Janusz Kacprzyk, Polish Academy of Sciences, Warsaw, Poland
e-mail: kacprzyk@ibspan.waw.pl

The series "Advances in Intelligent Systems and Computing" contains publications on theory, applications, and design methods of Intelligent Systems and Intelligent Computing. Virtually all disciplines such as engineering, natural sciences, computer and information science, ICT, economics, business, e-commerce, environment, healthcare, life science are covered. The list of topics spans all the areas of modern intelligent systems and computing such as: computational intelligence, soft computing including neural networks, fuzzy systems, evolutionary computing and the fusion of these paradigms, social intelligence, ambient intelligence, computational neuroscience, artificial life, virtual worlds and society, cognitive science and systems, Perception and Vision, DNA and immune based systems, self-organizing and adaptive systems, e-Learning and teaching, human-centered and human-centric computing, recommender systems, intelligent control, robotics and mechatronics including human-machine teaming, knowledge-based paradigms, learning paradigms, machine ethics, intelligent data analysis, knowledge management, intelligent agents, intelligent decision making and support, intelligent network security, trust management, interactive entertainment, Web intelligence and multimedia.

The publications within "Advances in Intelligent Systems and Computing" are primarily proceedings of important conferences, symposia and congresses. They cover significant recent developments in the field, both of a foundational and applicable character. An important characteristic feature of the series is the short publication time and world-wide distribution. This permits a rapid and broad dissemination of research results.

More information about this series at http://www.springer.com/series/11156

Jyotsna Kumar Mandal · Dhananjay Bhattacharyya
Nitin Auluck
Editors

Advanced Computing and Communication Technologies

Proceedings of the 11th ICACCT 2018

Springer

Editors
Jyotsna Kumar Mandal
Department of Computer Science and
    Engineering, Faculty of Engineering,
    Technology and Management
University of Kalyani
Kalyani, West Bengal
India

Nitin Auluck
Department of Computer Science
    and Engineering
Indian Institute of Technology Ropar
Rupnagar, Punjab
India

Dhananjay Bhattacharyya
Computational Science Division
Saha Institute of Nuclear Physics
Kolkata, West Bengal
India

ISSN 2194-5357   ISSN 2194-5365   (electronic)
Advances in Intelligent Systems and Computing
ISBN 978-981-13-0679-2   ISBN 978-981-13-0680-8   (eBook)
https://doi.org/10.1007/978-981-13-0680-8

Library of Congress Control Number: 2018941996

Printed on acid-free paper

This Springer imprint is published by the registered company Springer Nature Singapore Pte Ltd.
The registered company address is: 152 Beach Road, #21-01/04 Gateway East, Singapore 189721, Singapore

Preface

This AISC volume contains revised versions of the papers presented at the 11th International Conference on *Advanced Computing and Communication Technologies* (11th ICACCT 2018), held on 17–18 February 2018 at Asia Pacific Institute of Information Technology, Panipat, India.

The technical programme at the 11th ICACCT 2018 featured inaugural addresses by Conference General Chair Prof. Lalit M. Patnaik, National Institute of Advanced Studies, IISc, Bangalore; invited papers on 'Nonlinear Dynamics in Authentication'; 'Discrete Domain Swarm Robotics' and 'Microstrip Patch Antenna Array'; three contributed paper sessions and a valedictory session. The technical sessions were chaired by Prof. Jyotsna Kumar Mandal, Kalyani University; Prof. Dhananjay Bhattacharyya, Saha Institute of Nuclear Physics, Calcutta; Prof. Alok K. Rastogi, Institute for Excellence in Higher Education, Bhopal; Dr. Nitin Auluck, Indian Institute of Technology Ropar, Rupnagar; and Dr. Moirangthem Marjit Singh, NERIST, Nirjuli, Itanagar.

Out of 90 shortlisted papers for final review by the Technical Programme Committee and the expert panel, the final selection of 20 papers, corresponding to an acceptance rate of 20%, was selected by the evaluation processes subject to the acceptable similarity indices. The papers included cover a wide range of emerging research topics: spanning theory, systems, applications, critical reviews and case studies.

The volume includes state-of-the-art critical reviews on (i) chaos-based cryptography and authentication, (ii) gathering problem in discrete domain swarm robotics and (iii) wormhole attack in wireless sensor networks. The other chapters deal with emerging researches pertaining to bi-objective robust stochastic cellular facility layout; structural analysis in social networking, application of intuitionistic fuzzy sets in multi-criteria decision making, crowdfunding projects, audio repository for the visually impaired; deep learning framework for land-use/land-cover using remote sensing imagery; synchronization in NUMA multi-core processors;

neural network approach for classification of cardiovascular diseases; spectrum assignment in cognitive radio networks, energy-efficient IoT, minimizing the total range with two power levels in WSN, frequency selective defected structure for narrowband high Q applications, Schmitt Trigger Physical Unclonable Function, etc.

We would like to thank Prof. Nikhil Ranjan Pal, Indian Statistical Institute, Calcutta; Prof. Subir Sarkar, Jadavpur University, Calcutta; Prof. Partha P. Bhattacharya, Mody University of Science and Technology, Sikar, Rajasthan; Prof. Alok K. Rastogi, Institute for Excellence in Higher Education, Bhopal; Prof. Krishnendu Mukhopadhyaya, Indian Statistical Institute, Calcutta; Prof. Siddhartha Bhattacharyya, RCC Institute of Information Technology, Calcutta; Dr. Surya Prakash Singh, IIT Delhi; Dr. Srabani Mukhopadhyaya, BIT Mesra, Ranchi; Dr. Utpal Sharma, Tezpur University, Tezpur; Dr. Himadri Sekhar Dutta, Kalyani Government Engineering College, Kalyani; Dr. Mahavir Jhawar, Ashoka University, Sonepat; Dr. Tandra Pal, National Institute of Technology, Durgapur; and Dr. Moriangthem Marjit Singh, Northeast Regional Institute of Science and Technology, Itanagar, for their enormous contributions towards the success of this conference.

We are thankful to the authors of the finally accepted papers for suitably incorporating the changes suggested by the reviewers and the session chairs in the scheduled time frame.

Acceptance of the proposal submitted to Springer Nature Singapore Pte Ltd. for the publication of the proceedings is gratefully acknowledged. In this context, we are indebted to Mr. Aninda Bose, Senior Editor, Springer India Pvt. Ltd., New Delhi, for his valuable advices on similarity index and for instituting three Best Paper Presentation Awards worth EUR 200 each at the 11th ICACCT 2018. The selection for these awards was based on independent recommendations by the reviewers and session chairs.

We are grateful to the Institution of Electronics and Telecommunication Engineers (India) for technically sponsoring the conference since its inception in 2005.

The Organizing Committee patronized by the Managing Committee of Asia Pacific Institute of Information Technology SD India, viz. Mr. Vinod Gupta, Chairman; Mr. Umesh Aggarwal, Vice Chairman; Mr. Shrawan Mittal, Auditor, and guided by Prof. V. K. Shrivastava, Director, APIIT SD India, did a wonderful job in coordinating various activities. We would also like to thank Prof. Sachin Jasuja, Convener, ICACCT 2018, session's Coordinators, viz. Prof. Gaurav Gambhir, Dr. Sushil Sinha and Prof. Mahima Goel, all the faculty members and the administrative staff for their unreserved efforts to make the conference a great success. We would also like to thank the student volunteers, without their help and dedication the conference would not have been possible.

Finally, it is our pleasure to acknowledge the inspiring guidance of Conference General Chair, Prof. Lalit M. Patnaik, National Institute of Advanced Studies, Indian Institute of Science, Bangalore, in maintaining the high standards of the conference.

We hope the proceedings will inspire more research in *Advanced Computing and Communication Technologies*.

Kalyani, India   Jyotsna Kumar Mandal
Kolkata, India   Dhananjay Bhattacharyya
Rupnagar, India   Nitin Auluck

Conference Organization

Conference General Chair

Lalit M. Patnaik, INSA Senior Scientist and Adjunct Professor, Consciousness Studies Program, National Institute of Advanced Studies, IISc Campus, Bangalore-560012, India

Advisory Committee

Nikhil Ranjan Pal, Indian Statistical Institute, Calcutta, India
Pradeep K. Sinha, International Institute of Information Technology, Raipur, India
KTV Reddy, Institution of Electronics and Telecommunication Engineers, India
Subir Sarkar, Jadavpur University, Calcutta, India
Partha P. Bhattacharya, Mody University of Science and Technology, Sikar, India
Krishnendu Mukhopadhyaya, Indian Statistical Institute, Calcutta, India
N. Seetharamakrishna, Intel India Pvt. Ltd., Bangalore, India
P. N. Vinaychandran, Indian Institute of Science, Bangalore, India
Subhansu Bandopadhyay, University of Calcutta, Calcutta, India
Manish Bali, NVIDIA Graphics Pvt. Ltd., Bangalore, India
Cai Wen, Guangdong University of Technology, China
Yang Chunyan, Chinese Association of AI, China
Li Xingsen, Ningbo Institute of Technology, China
S. Venkatraman, University of Ballarat, Australia
Yashwant K. Malaiya, Colorado State University, USA
H. A. Nagarajaram, Centre of Computational Biology, CDFD, Hyderabad, India
Dinesh Khanduja, National Institute of Technology, Kurukshetra, India
Rachit Garg, Lovely Professional University, Phagwara, India
Sawtantar Singh, B. M. Singh College of Engineering, Muktsar, Punjab, India
Ekta Walia, South Asian University, New Delhi, India
E. G. Rajan, Pentagram Research Centre, Hyderabad, India
Pradosh K. Roy, Asia Pacific Institute of Information Technology, Panipat, India

K. Dasgupta, Kalyani Government Engineering College, India
K. R. Parpasani, MA National Institute of Technology, Bhopal, India
Kalyani Mali, University of Kalyani, Kalyani, India
Kanwalvir Singh Dhindsa, BBSB Engineering College, Punjab, India
M. Sandirigam, University of Peradeniya, Peradeniya, Sri Lanka
M. K. Bhowmik, Tripura University, Agartala, India
M. K. Naskar, Jadavpur University, Kolkata, India
Mahavir Jhawar, Ashoka University, Murthal, India
Md. U. Bokhari, Aligarh Muslim University, Aligarh, India
Meenakshi D'Souza, Indian Institute of Information Technology, Bangalore, India
N. R. Manna, North Bengal University, Siliguri, India
Nitin Auluck, Indian Institute of Technology Ropar, Rupnagar, India
P. Jha, Indian School of Mines, Dhanbad, India
P. K. Jana, North Bengal University, Siliguri, India
P. P. Sarkar, Purbanchal University, Koshi, Nepal
Paramartha Dutta, Visva Bharati University, Santiniketan, India
Partha S. Mandal, Indian Institute of Technology, Guwahati, India
Pitipong Yodmongkon, Chiang Mai University, Thailand
R. K. Jena, Institute of Management Technology, Nagpur, India
R. K. Samanta, North Bengal University, Siliguri, India
Rajendra Sahu, IIIT Hyderabad, India
Rajib K Das, University of Calcutta, Kolkata, India
Rameshwar Rijal, Kantipur Engineering College, Nepal
Rohit Kamal Chatterjee, BIT Mesra, Kolkata Extension Centre, Kolkata, India
S. Dutta, B. C. Roy Engineering College, Durgapur, India
S. Mal, Kalyani Government Engineering College, India
S. Mukherjee, Burdwan University, Burdwan, India
S. Muttoo, Delhi University, Delhi, India
S. Shakya, Tribhuvan University, Nepal
S. K. Mondal, Kalyani Government Engineering College, India
Sarmishtha Neogy, Jadavpur University, Kolkata, India
Sergei Silvestrov, Mälardalen University, Sweden
Shuang Cang, Bournemouth University, UK
Siddhartha Bhattacharyya, RCC Institute of Information Technology, Calcutta, India
Srabani Mukhopadhyaya, BIT Mesra, Lalpur Extension Centre, Ranchi, India
Subhamoy Changder, National Institute of Technology, Durgapur, India
Sudip Roy, Shell Technology Centre, Bangalore, India
Surya P. Singh, Indian Institute of Technology, New Delhi, India
Szabo Zoltan, Corvinus University of Budapest, Hungary
Tandra Pal, National Institute of Technology, Durgapur, India
Teresa Goncalves, Universidade de Évora, Portugal
Trupti R. Lenka, National Institute of Technology, Silchar, India
Ujjal Maulik, Jadavpur University, Calcutta, India
Utpal Sharma, Tezpur University, Tezpur, India

Reviewers

Akash K. Tayal, IG Delhi Technical University for Women, New Delhi, India
Alok K. Rastogi, Institute for Excellence in Higher Education, Bhopal, India
Ambar Dutta, BIT Mesra, Kolkata Extension Centre, Kolkata, India
B. K. Tripathy, Vellore Institute of Technology, Vellore, India
Bijay Baran Pal, University of Kalyani, Kalyani, India
Buddhadeb Sau, Jadavpur University, Calcutta, India
Dilip K. Pratihar, Indian Institute of Technology, Kharagpur, India
Debasish Jana, TEOCO Software Pvt. Ltd., Kolkata, India
Deepanwita Das, National Institute of Technology, Durgapur, India
Dhananjay Bhattacharyya, Saha Institute of Nuclear Physics, Calcutta, India
Himadri Sekhar Dutta, Kalyani Government Engineering College, Kalyani, India
Jyotsna Kumar Mandal, University of Kalyani, Kalyani, India
Krishnendu Mukhopadhyaya, Indian Statistical Institute, Calcutta, India
Mahavir Jhawar, Ashoka University, Sonepat, India
Moirangthem Marjit Singh, NERIST, Itanagar, India
Nitin Auluck, Indian Institute of Technology Ropar, Rupnagar, India
Paramartha Dutta, Visva Bharati University, Santiniketan, India
Partha P. Bhattacharya, Mody University of Science and Technology, Sikar, India
Partha Sarathi Mandal, Indian Institute of Technology, Guwahati, India
Pradosh K. Roy, Asia Pacific Institute of Information Technology, Panipat, India
R. K. Amit, Indian Institute of Technology, Madras, India
R. R. K. Sharma, Indian Institute of Technology, Kanpur, India
Rajib K. Das, University of Calcutta, Kolkata, India
Rohit Kamal Chatterjee, BIT Mesra, Kolkata Extension Centre, Kolkata, India
Siddhartha Bhattacharyya, RCC Institute of Information Technology, Calcutta, India
Somnath Mukherjee, Calcutta Business School, Calcutta, India
Srabani Mukhopadhyaya, Birla Institute of Technology, Mesra, Ranchi, India
Subhamoy Changder, National Institute of Technology, Durgapur, India
Subhas Bhagat, Indian Statistical Institute, Calcutta, India
Subho Chaudhuri, BIT Mesra, Kolkata Extension Centre, Kolkata, India
Subir Sarkar, Jadavpur University, Calcutta, India
Sumer Singh, Indian Institute of Technology, Delhi, India
Sunandan Sen, University of Calcutta, India
Surya P. Singh, Indian Institute of Technology, New Delhi, India
Tandra Pal, National Institute of Technology, Durgapur, India
Ujjal Maulik, Jadavpur University, Calcutta, India
Utpal Sharma, Tezpur University, Tezpur

Organizing Committee

Patrons

Vinod Gupta

Chairman, Asia Pacific Institute of Information Technology, Panipat

Umesh Aggarwal

Vice Chairman, Asia Pacific Institute of Information Technology, Panipat

Sharwan Mittal

Auditor, Asia Pacific Institute of Information Technology, Panipat

Convener

Sachin Jasuja, Department of Mechatronics and EE, APIIT, Panipat

Session Coordinators

Gaurav Gambhir, Department of CSE, APIIT, Panipat
Mahima Goel, Department of Electronics Engineering, APIIT, Panipat
Sushil Kumar Sinha, Department of Mathematics, APIIT, Panipat
Sanjeev Sharma (Technical Support)

Website Management

Sachin Jain, Multimedia Management Unit, APIIT, Panipat

Publication

Ravi Sachdeva, Department of CSE, APIIT, Panipat

Sponsorship and Finance

Prateek Mishra, Department of CSE, APIIT, Panipat
Sanjeev Jawa, Finance and Accounts Section, APIIT, Panipat

Registration

Virender Mehla, Department of Electronics Engineering, APIIT, Panipat
Priyanka Sachdeva, Department of CSE, APIIT, Panipat
Ginni Chawla, Department of Electronics Engineering, APIIT, Panipat

Hospitality, Transport, Accommodation

Pardeep Singla, Department of Electronics Engineering, APIIT, Panipat
Anuj Gupta, Department of Electronics Engineering, APIIT, Panipat
Manoj Kumar, Department of Management, APIIT, Panipat
Anil Jaukhani, Asstt. Administrative Executive, APIIT, Panipat
Rajinder Khurana, Material Management Unit, APIIT, Panipat

Message from the General Chair

It is gratifying to note that the Asia Pacific Institute of Information Technology organized the '11th International Conference on Advanced Computing and Communication Technologies' (11th ICACCT^{TM} 2018) during 17–18 February 2018 at the historical surrounding of Panipat, Haryana.

High-quality research contributing to the fields of 'Advanced Computing and Communication Technologies' on a wide range of research topics: spanning theory, systems and applications were solicited for presentation at the conference. Out of the contributed papers, 90 papers were shortlisted for rigorous review by the expert panel after content evaluation and anti-plagiarism check using Turnitin$^{®}$. The final selection of 20 papers corresponding to an acceptance rate of 20% was determined by these evaluation processes. The post-conference publication of the proceedings has facilitated revisions by the authors presenting their papers, considering the comments and suggestions from the reviewers, session chairs and the audience. We had parallel sessions on 17 February 2018 covering a wide range of emerging research topics, e.g. machine learning, facility layout problem, crowdfunding projects, deep learning, MHD nanofluid flow, medical diagnostics, human–computer interface, social networking, system performance, wireless sensor networks, cognitive radio networks, antenna design.

Invited papers by Prof. Jyotsna Kumar Mandal, Prof. Alok K. Rastogi and Dr. Deepanwita Das et al. on 'Nonlinear Dynamics in Authentication', 'Microstrip Patch Antenna' and 'Discrete Domain Swarm Robotics' are also included in the proceedings.

We are grateful to Springer Nature Singapore Pte Ltd., Singapore, and the Institution of Electronics and Telecommunication Engineers, New Delhi, India, for being the publication and technical sponsors. The three Best Paper Presentation Awards of EUR 200 each, instituted by Springer India Pvt. Ltd., New Delhi, for the 11th ICACCT 2018 are gratefully acknowledged too. My special appreciation to Mr. Aninda Bose, Senior Editor, Springer India Pvt. Ltd., New Delhi, for his continued support and proactive role in promoting ICACCT series.

As General Chair of the conference, I would like to place on record my profuse thanks to Prof. Nikhil Ranjan Pal, Indian Statistical Institute, Calcutta, India;

Prof. Partha P. Bhattacharya, Mody University of Science and Technology, Sikar, Rajasthan; Prof. Krishnendu Mukhopadhyaya, Indian Statistical Institute, Kolkata; Dr. Dilip K. Pratihar, IIT Kharagpur; Prof. Siddhartha Bhattacharyya, RCC Institute of Information Technology, Calcutta; Dr. Rajib K. Das, University of Calcutta; Dr. Ambar Dutta, BIT Mesra, Kolkata Extension Centre; Dr. Rohit K. Chatterjee, BIT Mesra, Kolkata Extension Centre; Dr. Partha Sarathi Mondal, IIT, Guwahati; Dr. Utpal Sharma, Tezpur University; Dr. Moirangthem Marjit Singh, NERIST, Itanagar, for their enormous contributions towards the success of this conference.

The members of the Editorial Committee, viz. Prof. Jyotsna Kumar Mandal, University of Kalyani; Prof. Dhananjay Bhattacharyya, Saha Institute of Nuclear Physics, Calcutta, and Dr. Nitin Auluck, IIT Ropar; the Programme Chair and Co-chairs, viz. Prof. V. K. Shrivastava, APIIT, Panipat; Dr. Surya Prakash Singh, IIT Delhi, and Dr. Srabani Mukhopadhyaya, BIT Mesra, Ranchi, deserve our sincere appreciations.

My sincere thanks also to the session chairs, reviewers, the organizing committee members and the student volunteers.

In addition, our sincere thanks to Prof. Pradosh K. Roy who did the lion's share of the editing of these proceedings.

Finally, I would like to thank the APIIT management for providing all the generous and excellent support in promoting the conference. Special words of thanks go to Prof. Sachin Jasuja, Prof. Gaurav Gambhir, Prof. Ravi Sachdeva, Dr. Sushil K. Sinha, Prof. Mahima Goel, all the faculty members and the administrative staff of APIIT for their overwhelming support at every stage of the organization of the conference.

I hope the proceedings would be a valuable addition in the technical literature on computing and communication technologies.

Bangalore, India   Lalit M. Patnaik
March 2018   INSA Senior Scientist and Adjunct Professor
                                                       Consciousness Studies Program
                                    NIAS, Indian Institute of Science, Bangalore, India
                                                                    www.lmpatnaik.in

Contents

About the Editors

Jyotsna Kumar Mandal is former dean of the Faculty of Engineering, Technology and Management and a senior professor at the Department of Computer Science and Engineering, University of Kalyani, India. He obtained his Ph.D. (engineering) from Jadavpur University, India. He has co-authored six books: Algorithmic Design of Compression Schemes and Correction Techniques—A Practical Approach; Symmetric Encryption—Algorithm, Analysis and Applications: Low Cost-based Security; Steganographic Techniques and Application in Document Authentication—An Algorithmic Approach; Optimization-based Filtering of Random Valued Impulses—An Algorithmic Approach; Artificial Neural Network Guided Secured Communication Techniques: A Practical Approach and Handbook of Research on Natural Computing for Optimization Problems (two volumes). He has authored more than 350 papers on a wide range of topics in international journals and proceedings. His areas of research include coding theory, data and network security, remote sensing and geographic information system (GIS)-based applications, data compression, error correction, visual cryptography and steganography, and distributed and shared memory parallel programming. He is a fellow of the Institution of Electronics and Telecommunication Engineers and a member of IEEE, ACM, CRSI and the Computer Society of India.

Dhananjay Bhattacharyya is head of the Computational Science (CS) Division, Saha Institute of Nuclear Physics, Calcutta, India. He obtained his Ph.D. from the Indian Institute of Science, Bangalore, and a postdoctoral fellowship from National Institutes of Health, Bethesda, Maryland, USA. His research interest is in understanding the structure–function relationship of biological macromolecules, particularly nucleic acids. In order to understand different structural features of nucleic acids, the CS Division at the Saha Institute of Nuclear Physics has developed a number of software tools under his guidance, such as NUPARM, BPFIND and PyrHB Find. Using some of these tools, the CS Division has classified the structures of different non-canonical base pairs appearing in RNA crystal structures. He has successfully guided several doctoral candidates on computational biology and bioinformatics.

Nitin Auluck is associate professor in the Department of Computer Science and Engineering at the Indian Institute of Technology Ropar, Rupnagar, Punjab, India. Prior to that, he was an assistant professor at the Department of Computer Science at Quincy University, Illinois, USA. He obtained his Ph.D. in computer science and engineering from the University of Cincinnati, Cincinnati, Ohio, USA, in 2005. He has published in IEEE Transactions on Parallel and Distributed Systems (PDS), Transactions on Computers and Intelligent Systems, Proceedings of the IEEE International Symposium on Parallel Architectures, Algorithms and Programming 2012, Taipei, Taiwan, etc. His current areas of research are real-time systems, scheduling and parallel and distributed computing.

Part I
Advanced Computing

Authentication Based on Nonlinear Dynamics

Jyotsna Kumar Mandal

Abstract This paper considers a nonlinear logistic map to generate random real number from where binary random sequence has been generated. Process of encryption and decryption has also been described. A genetic algorithm-based search results are shown in graphical manner to illustrate the generated optimized seed as system parameter and initial value of the variable.

Keywords Steganography · Nonlinear dynamics · PN sequence
Logistic maps · Genetic algorithm

1 Introduction

Chaos is an omnipresent event active in deterministic nonlinear systems which show extreme sensitivity to initial conditions and have haphazard-like behaviors. Chaotic behavior was discovered by Edward N. Lorenz in 1963 in the context of atmospheric convection, albeit it can be traced back to the works of the polymath Henri Poincaré (1854–1912). Actually, chaos theory is a field of study in mathematics which has applications in several disciplines like Physics, Engineering, Biology, and philosophy.

Let us define a disconnected dynamical scheme in the general form:

$$z_{n+1} = f(z_n), f : I \to I, z_0 \in I \tag{1}$$

where f is a unremitting plot in the interval $I = [0, 1]$.

The process may be called chaotic if some conditions are fulfilled. Like it is receptive to initial conditions but generates repeating identical sequence for the same initial conditions. There exists topological transitivity, and density periodic points are identical for identical periodicity but extremely nonperiodic for nonho-

J. K. Mandal (✉)
Department of Computer Science and Engineering, Faculty of Engineering,
Technology and Management, University of Kalyani, Kalyani, West Bengal, India
e-mail: jkmandal@klyuniv.ac.in; jkm.cse@gmail.com

© Springer Nature Singapore Pte Ltd. 2019   3
J. K. Mandal et al. (eds.), *Advanced Computing and Communication Technologies*,
Advances in Intelligent Systems and Computing 702,
https://doi.org/10.1007/978-981-13-0680-8_1

mogeneous points. There are pseudo-random property exists in chaotic maps. Equation 1 shows a linear chaotic map. Some good properties of chaos are topological transitivity, nonperiodicity, and ergodicity. Ergodicity generates good amount of confusion in cryptographic process but the output has the same allotment for any input. The maps are very much sensitive to initial conditions/control parameter which means it generates good amount of dispersal with a small alteration in the plaintext or secret key as a result a very small amount of variation in the input it generates enormous change in the output values. This nonlinear system has very good mixing property which means during encryption, diffusion with a small change in one plain block of the whole plaintext produces different set across entire output space. A small deviation in the local area can generate a huge change in the whole space. The chaos follows deterministic dynamics and generates deterministic pseudo-randomness which means the process can cause a random-like behavior. The structural complexity of the nonlinear dynamics leads to algorithmic complexity which means an uncomplicated progression has a very high complexity.

In this paper, generation of sequence of real numbers for a given system parameter and starting value using a liner logistic map has been discussed. Generation of pseudo-random number from these real numbers is given along with the sequence of testing of number for true randomness. Use of these system parameters, starting value of the variable for authentication of images is given. Both spatial and transform domain fabrication of authenticating bits at the sender side are described. Also authentication process at the destination for both domains has also been elaborated.

Section 2 of the paper fabricates the related study of the matter. Section 3 deals with proposal for embedding, and authentication both in spatial and transform domain of cover image is given extensively. Outcomes are discussed in Sect. 4 along with analysis. Wrapping up is given in Sect. 5, and the references at end.

2 Review Work

Robert Matthews used logistic map in binary images in 1989 [1]. S. Som. and A. Kotal used 1D logistic map through Arnold's cat map in grayscale images [2] in 2012. Jiri Fridrich applied 2D baker map [3] in gray images in 1998. Sukalyan Som and Sayani Sen used cross-coupled chaotic tent map [4] in grayscale Images in 2013. Logistic map in RGB color images has been used by N. K. Pareek et al. in 2006. Plaintext is used by N. K. Pareek et al. in multiple one-dimensional chaotic map [5] in 2005. Arnold's cat map is used in 1D logistic map [6] for DNA coding by L. Liu, Q. Zhang, and X. Wei in 2012. RGB color image was used in logistic map by Xingyuan et al. in 2012 [7]. Seyed Mohammad Seyedzadeh and Sattar Mirzakuchaki used coupled two-dimensional piecewise nonlinear chaotic map for encoding RGB [8] color images in 2012. Cross-coupled chaotic tent map was used by Haojiang Gao et al in 2006 for map-based bit generator (CCCBG) [9]. Xu Shu-Jiang et al. used logistic map in Lorenz system for grayscale image encoding.

Alireza et al. used baker's map to encode grayscale image using simplified AES in 2011. Plaintexts were encoded by Mina Mishra et al. in 2011 using sinusoidal nonlinear shift register and logistic map.

In this paper, a linear logistic map is used to generate real numbers using predetermined system constraints and opening value of the logistic variable. Binary stream is generated from this sequence. The encoding is done using generated bit stream both in spatial and transform domain.

3   Proposal with Results

The proposal consists of two sections. The first section considers a linear logistic map (as given in Eq. 1) from where seeds are chosen manually to obtain a sensitive region of the chaotic equation along with initial value and twenty real numbers are generated. The mean of these twenty numbers are obtained, and this mean is designated as threshold value. Twenty binary bits are encoded based on this threshold from generated real numbers. These numbers are tested for monobit, serial, and Poker test to satisfy the pseudo-randomness of the PN sequence. In the next section, these bit streams are embedded into twenty LSB pixels of an image (say) to illustrate the embedding process of generated chaotic bit sequence generated from chaotic map.

$$z_{n+1} = ¥ \, z_n \, (1 - z_n) \tag{2}$$

where $0 <= ¥ <= 4$, $0 < z_n < 1$ and the chaotic region is $3.5699456 < ¥ <= 4$, $¥$ is control parameter. z_n: $n = 0, 1, 2, \ldots$ generates nonperiodic and nonconvergent real number. Table 1 shows the real numbers generated using the value of the system parameter as $¥ = 3.55$ and $z_0 = 0.519$. The mean of these twenty numbers is generated as $M = 0.634529$ which is threshold $(M = T)$, and the binary stream is generated using following Eq. 3 (Fig. 1).

$$\begin{aligned} G(z_{n+1}) &= 0 \quad \text{if } z_{n+1} = T \\ &= 1 \quad \text{if } z_{n+1} \, T \end{aligned} \tag{3}$$

In steganographic application, for embedding, let us consider a byte of bits generated from chaotic equation taken from left to right as $B_k = 10101010$. Take eight bits from secret image. XOR with first bit of Chaotic Sequence is done to generate the encoded secret byte. Let one such byte is $C_k = 01010000$. First bit of chaotic sequence B_k is 1: Insert seven 1 toward MSB of B_k. $B_k = 11111111$. Make XOR between B_k and C_k. $C_k = 01010000$, $B_k = 11111111$, $C'_k = 10101111$. Insert these bits into eight bytes of the cover image. Let eight bytes of cover image are 1011010: 11010101: 11001010: 01101100: 01110011: 10101010: 00101100: 00011101: Consider LSB embedding. On embedding $C'_k = 10101111$. These embedded eight

Table 1 Real numbers generated from Eq. 2 for ¥ = 3.55 and z_0 = 0.519, threshold T = 0.634529

Real numbers (Z_{n+1})	0.886218	0.357965	0.815883	0.533274	0.88357	0.365204	0.822997	0.51714	0.886457	0.357311
Binary numbers (G)	1	0	1	0	1	0	1	0	1	0
Real numbers (Z_{n+1})	815221	0.534756	0.883212	0.366179	0.823926	0.515005	0.886701	0.356642	0.814542	0.536274
Binary numbers (G)	1	0	1	0	1	0	1	0	1	0

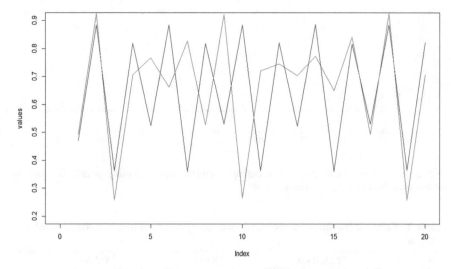

Fig. 1 Sensitivity of chaotic equations

bytes are 1011011, 11010100, 11001011, 01101100, 01110011, 10101011, 00101101, and 00011101. This may be termed as embedded image.

In the receiving end, same logistic map, same system parameter, and the initial values are used to generate twenty binary numbers. Consider first eight generated number is 10101010. Take first bit from chaotic sequence. First bit of chaotic sequence is B_n is 1: Insert seven 1 toward MSB of B_n. $B_n = 11111111$; make XOR between B_n and C'_k, $B_n = 11111111$, $C'_k = 10101111$, $C_k = 01010000$: This is the secret information.

After embedding, we use adjustment of the pixel without hampering the embedded bit to enhance the quality of image. Genetic algorithm is used to optimize the pixel value toward near equal to the original pixel value before embedding without altering the embedded bits. In that case, some other embedding techniques may be used along with hash function to select the embedding position.

For optimization of seeds and system parameter genetic algorithm is used and generation wise progress has been shown in Fig. 2 where multi-object (system parameter and initial value) is optimized through use of 100 generation through genetic algorithm. In this case, the optimized value of system parameter is 3.91660377519162 and the initial condition is 0.0464217819826947.

Table 2 shows the chi-square values obtained based on monobit, serial, and Poker test on optimized seed obtained from genetic algorithm which shows that all three tests are satisfied when we use the optimized seed and the system parameter.

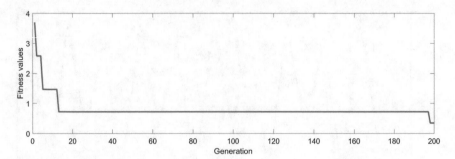

Fig. 2 Generation-wise progress curve during optimization process using genetic algorithm to obtain optimized seed and system parameter

Table 2 Values of chi-square on optimized seed and system parameter in monobit, serial, and Poker test

Test	Monobit	Serial	Poker
Value	0.500	3.4577	0.3488

4 Conclusions

In this article, the use of chaotic maps and logistic equation is described for generation of random binary number along with the test of randomness. Sample process of embedding and extraction has also been described. Searching of optimized seed and system parameter is also shown using genetic algorithm.

References

1. Matthews, R.: On the derivation of a "chaotic" encryption algorithm. Cryptologia **13**(1), 29–41 (1984)
2. Som, S., Kotal, A., Chatterjee, A., Dey, S., Palit, S.: Confusion and diffusion of color images with multiple chaotic maps and chaos-based pseudorandom binary number generator. In Proceedings 1st International Conference on Emerging Trends and Applications in Computer Science (ICETACS 2013), pp. 1–5 (2013)
3. Fridrich, J.: Symmetric ciphers based on two-dimensional chaotic maps. Int. J. Bifurc. chaos **8**(06), 1259–1284 (1998)
4. Som, S., Sen, S.: A non-adaptive partial encryption of grayscale images based on chaos. In First International Conference on Computational Intelligence: Modelling, Techniques and Applications (CIMTA-2013), Procedia Technology, vol. 10, pp. 663–671 (2013)
5. Pareek, N.K., Patidar, V., Sud, K.K.: Image encryption using chaotic logistic map. Image Vis. Comput. **24**(9), 926–934 (2008)
6. Pareek, N.K., Patidar, V., Sud, K.K.: Cryptography using multiple one-dimensional chaotic maps. Commun. Nonlinear Sci. Numer. Simul. **10**(7), 715–723 (2012)

7. Wang, X., Teng, L., Qin, X.: A novel colour image encryption algorithm based on chaos. Signal Processing **92**(4), 1101–1108, (2012). [https://doi.org/10.1016/j.sigpro.2011.10.023]
8. Pareek, N.K., Patidar, V., Sud, K.K.: Discrete chaotic cryptography using external key. Phys. Lett. A **309**(1–2), 75–82 (2003)
9. Gao, H., Zhang, Y., Liang, S., Li, D.: Cross-coupled chaotic tent. Map Based Bit Generator (CCCBG) (2010)

Gathering in the Discrete Domain: State of the Art

Arun Sadhu, Madhumita Sardar, Deepanwita Das
and Srabani Mukhopadhyaya

Abstract Discrete domain swarm robotics is an emerging and challenging field of research. Unlike continuous domain where working place of the robots is a two-dimensional plane, in the discrete domain their working place is modelled by a graph. Robots are deployed on the nodes of a given graph, and they are allowed to move only along the edges of that graph. Consequently, the models used in the continuous domain are not always applicable in the discrete domain. Exemplified by the gathering problem, this article critically reviews models, assumptions, and approaches that have been proposed in solving the problems in the discrete domain.

Keywords Robot swarm · Gathering · Discrete domain

1 Introduction

Robots are self-controlled machines that are designed to perform various tasks to reduce the workload of human being in many areas such as car body assembling and painting, space explorations and heavy traffic control. Moreover, in some domain of applications, they are absolutely required to protect fatal life loss such as automated demining, radioactive matter search and fire fighting. A single efficient programmable robot is desirable in many applications, whereas using small

A. Sadhu
Dr. B. C. Roy Engineering College, Durgapur 713206, India
e-mail: arun.sadhu@bcrec.ac.in

M. Sardar · D. Das (✉)
National Institute of Technology Durgapur, Durgapur 713209, India
e-mail: deepanwita@cse.nitdgp.ac.in

M. Sardar
e-mail: ms.17cs1108@phd.nitdgp.ac.in

S. Mukhopadhyaya
Birla Institute of Technology Mesra, Lalpur Extension Centre, Ranchi 834001, India
e-mail: smukhopadhyaya@bitmesra.ac.in

© Springer Nature Singapore Pte Ltd. 2019   11
J. K. Mandal et al. (eds.), *Advanced Computing and Communication Technologies*,
Advances in Intelligent Systems and Computing 702,
https://doi.org/10.1007/978-981-13-0680-8_2

multiple robots may produce similar results while reducing cost of production. Recently, the idea of using a group of robots (known as swarm robots) instead of single complex robot has emerged. Robot swarm is inspired by the behaviour of natural entities such as ants, bees and birds which collaboratively perform complex tasks. The two factors that have enhanced the use of swarm robots are their ability of collaboration in work and cost-effective nature with simple design.

In most of the research works, the robots in a swarm are represented as points on a two-dimensional Euclidean space. Some researchers instead have depicted robots as unit discs, calling them as fat robots. Though many problems have been solved in the continuous domain using the swarm, researchers in the recent years have tried to discretize the plane. In the discrete domain, the working space is modelled by a graph; the robots supposed to be deployed over the nodes of a graph and are allowed to move only along the edges of the graph. The node-to-node movement of a robot is instantaneous which ensures that the robots can always be seen on the nodes and never on the edges. Finding out the minimal set of features to achieve optimum performance by the robots is the basic objective of this field of research. Like continuous domain, here also various challenges occur due to consideration of different models and limited features of the robots.

Depending on features of discrete domain, problem definition as well as nature of the solution varies. Using gathering problem in discrete domain as the running example, this paper tries to analyse different approaches of problem solving in discrete domain. In Sect. 1, the gathering problem is defined followed by various aspects of the problem. A detailed discussion on state of the art of gathering in discrete domain is presented in Sect. 2. Finally, Sect. 3 concludes the paper with pointers to the future scope of research.

2 Gathering Problem: State of the Art

Fraigniaud et al. [10] have addressed rendezvous problem, the gathering of two robots/agents, in an anonymous tree with locally labelled ports. The two robots are considered as identical, synchronous, abstract state machines with a little amount of memory. Gathering or meeting of these two robots is impossible if the initial positions of the robots are symmetrical. The robots explore the tree first to compute the maximum degree of the tree. Next they search in phases, exploring the tree partially in each sub-phase. If the robots meet during the searching phase, they perceive the fact that they have already gathered and terminate the process at that point. It is shown that rendezvous problem can be solved using $O(\log n)$ bits of memory in a tree of size at most n.

Using the similar concept of abstract state machines and port labelling, Shibata et al. [15] have presented a variation of gathering called g-partial gathering in trees where robots gather at different nodes in such a way that at least g robots meet at any node. Taking multiplicity detection as a key factor, this problem is solved in three models: (a) *weak multiplicity detection* with no token, where partial gathering is achieved in $O(kn)$ moves in asymmetrical trees where n is the number of nodes and k is the number of robots, (b) *strong multiplicity detection* with no token where the robots gather at the centre node of the tree. In case of two centre nodes, the robots move in such a way so that at least g agents meet at each of the centre nodes and (c) *weak multiplicity detection* with removable tokens where the robots elect some leaders. These leaders then instruct other robots where to move in order to gather.

To overcome the drawbacks of the previous two works where robots required memory, multiplicity detection capability and tokens to mark nodes, D'Angelo et al. [6] have proposed a solution to gather anonymous, asynchronous and oblivious robots with unlimited visibility on a tree where robots follow the CORDA model and do not have multiplicity detection capability. Here, robots gather at the centre of the tree. In case of two centres, if the two sub-trees rooted at those two centres are isomorphic, then gathering is impossible even with multiplicity detection capability. If not, then the root of the sub-tree with lesser number of occupied nodes will be the gathering point. When the two sub-trees are of the same size, the root of the sub-tree with the robots closer to the root is taken as the gathering point.

In the previous work [6], the topology of the tree was exploited as the robots could view the entire tree due to unlimited visibility. Removing the assumption of unlimited visibility makes the problem harder to solve. Bhaumik and Chaudhuri [2] have considered limited visibility of the robots and relaxed the problem environment by uniquely marking one of the nodes of the tree as the target node for gathering. The robots are anonymous, oblivious, asynchronous and are equipped with weak multiplicity detection capability. They traverse the tree randomly looking for the target node until they find it. If a robot reaches a leaf, it takes a 180° turn and keeps marking the visited vertices until it finds another open path. The marking is done in order to close down paths that do not lead to the target node. This approach requires marking the visited vertices that requires memory. However, it is not mentioned in the paper whether the memory resides on the nodes or on the edges. Also the idea of robots taking a 180° turn does not comply with the notion of a graph.

Sinha and Mukhopadhyaya [16] have proposed another solution to the same problem where they have considered dedicated memory to the nodes, known as whiteboards. They have proposed a solution for target node searching in trees, which is devoid of the issues present in the last approach. It is assumed that the target node is visible to at least one robot which initiates the process by marking its whiteboard as 1. Once a neighbour node of any robot gets its whiteboard marked, the robot also activates itself by marking its own whiteboard one greater than that of the mark of its neighbour node. In case of more than one neighbours, the marking is estimated by considering the node with lowest mark. Every robot is responsible to explore the sub-tree rooted at any of its unmarked, unoccupied neighbour. When a robot finds that all its adjacent nodes are marked or occupied, it starts moving towards the target node by following a path along the smaller whiteboard marks. When the robots are fully synchronous, the algorithm takes $O(n)$ computational cycles, where n is the number of nodes in the tree. It also guarantees a complete exploration of the tree.

These approaches vary with the topologies considered. Considering the topology as grid, D'Angelo et al. [6] have discussed every possible gatherable configuration along with a list of impossibility situations. A grid configuration is ungatherable if it is periodic on a grid with at least one even dimension or if it is symmetrical with the axis of symmetry passing through edges. If a configuration is invariant with respect to rotations of 90 or 180°, where the geometric centre of the grid coincides with the rotation point, then the configuration is said to be periodic. For all the other cases, multiplicity detection is not required for gathering in grids by anonymous, asynchronous, oblivious robots equipped with unlimited visibility except in a [2 × 2] grid. They have proposed the gathering points based on three types of grids. **(a)** In [odd × odd] grids, robots gather at the central node of the grid while in **(b)** [odd × even] grids, one of the two central nodes of the odd borders of the grid is the gathering point. The grid is divided into two halves from the even side, and the central node of the odd border of the half having more number of robots is chosen as the gathering point. If the two halves have equal number of robots, then four binary strings are generated from the four corner vertices, each of them representing the configuration of the grid. Each binary string is generated starting from each corner and proceeding in the direction parallel to the line joining the two central nodes of the odd borders. The central node of the odd border where the corner generating the lexicographically largest string resides is chosen as the gathering node. **(c)** Similarly for [even × even] grids, a sequence showing the number of empty nodes between occupied nodes in the grid is constructed starting from each corner. The corner providing the minimal sequence is the gathering point.

Gathering in grids with limited visibility is a challenge as the gathering points cannot be decided in the ways discussed in [6]. Absho et al. [1] have gathered a closed chain of robots on a grid. The connectivity of the chain is maintained throughout the process. The robots are autonomous, synchronous and follow CORDA model. The robots can only see the sub-chain that contains the robot itself and its next few chain neighbours in both the directions. Gathering is performed by shortening the chain by merging. Merging implies whenever a robot hops into an already occupied position, the two robots are merged and treated as one. In cases where merging is not possible, reshaping of the chain is done. In the solution by Cord-Landwehr et al. [3], the robots do not form a chain. Assuming that the swarm is connected, the authors consider the outer boundary of the swarm as a closed chain with fringes (if any). Gathering is achieved by shortening this outer boundary using merging or reshaping the boundary.

Even though both the above works [1, 3] have optimal running times of $O(n)$, n being the number of robots, the robots require to maintain a constant number of states, constant amount of memory and local communication. Fischer et al. [9] have eliminated these overheads and designed hop patterns for the boundary robots such that the algorithm's performance depends quadratically on the length of the outer boundary of the swarm area. The drawback of the strategies in [1, 3, 9] is that instead of gathering on a single point, they have gathered the robots within a 2 × 2 sub-grid.

Di Stefano and Navarra [7] have considered infinite grids. An infinite grid does not contain the corner vertices as there are no borders. The robots are oblivious, asynchronous, without communication, have unlimited visibility as well as multiplicity detection capability and follow the CORDA model. Here, the concept of Weber point has been used to compute the gathering point, and hence, the approach is optimum with respect to the total number of moves. Dutta and Chaudhuri [8] have eliminated the assumptions of unlimited visibility and multiplicity detection capability. However, the robots are full-compass which makes the problem easier. It is assumed that the visibility graph is acyclic and remains connected throughout the process. The moves of the robots are designed in such a way that only a robot with degree one in the visibility graph can move towards its immediate neighbour and merge with it. Hence, there always exists a robot which is eligible to move till all the robots gather. While merging with its neighbour, the robot does not create any new edge connection in the visibility graph. The number of nodes in the visibility graph decreases when the robots merge and eventually become one.

Guilbault and Pelc [11] have taken a regular bipartite graph as working space and considered gathering with robots which are identical, oblivious, asynchronous, have limited visibility and multiplicity detection capability and follow CORDA model. It has been proved that the only initial deployment configuration which supports a successful gathering is a star graph of size at least 3. The algorithm allows the robots with one occupied neighbour to move. In a star, there is exactly one robot with more than one occupied neighbour which cannot move. All other singleton robots are adjacent to it. The robots at singletons move towards the centre occupied by this stable robot and stop.

Rajsbaum et al. [14] present a very different model from the ones discussed above. All robots need not to be present at the graph initially. They can appear at any time during the execution. The robots are asynchronous, follow CORDA model, can communicate with lights and can suffer from crash faults. Since these assumptions make gathering difficult, the gathering problem is relaxed by introducing the edge gathering problem, where instead of gathering on one vertex, robots can gather on the end vertices of an edge or the 1-gathering problem, where robots can gather on the vertices of a complete sub-graph. The ungatherable situations are also listed out in this paper.

Gathering of robots on an unoriented anonymous ring is also providing us a great motivation due to the existence of possible symmetries, when robots are deployed on the ring [13]. Robots, deployed on an unoriented ring, cannot differentiate between clockwise and anticlockwise direction and therefore cannot exploit the topological structure of a ring. However, a ring with any predetermined orientation can give an easy solution to the problem like target searching [16] on that ring.

Klasing et al. [12] first came with the idea of gathering of asynchronous robots on a ring. In their paper, they have introduced the concept of viewing configuration which consists of two subsets, one subset contain numbers indicating distances between two consecutive robot positions and the other subset stores the record of multiplicity points on the ring looking towards clockwise direction from itself to be

given a feasibility study for all non-gatherable configurations and algorithms for all gatherable configurations.

Robots here considered with unlimited visibility and follow the CORDA model. From the initial configuration, robots check for the multiplicity points on the ring. When only one multiplicity point is present, robots on the non-multiplicity point try to reach there by the shortest path. In case no multiplicity point exists on the ring, the configuration is gatherable if a leader-like robot can be elected. If it is possible to elect a robot as a leader, then the leader moves to any one of its occupied neighbours to create a single multiplicity point. They have shown that for presence of an odd number of robots on a ring, gathering is feasible for all non-periodic configurations. The authors also listed some impossibility results in their paper, such as, presence of an odd number of robots on a ring having any periodic initial configuration would not lead to a successful gathering. Gathering is also not possible in case of even number of robots having *edge–edge* symmetry.

Klasing et al. [13] extended the study by showing influence of symmetry on the initial configuration of the robots during gathering. They have proposed symmetry-preserving method for more than 18 robots starting from all initial configurations for which gathering is feasible, provided ring has only one axis of symmetry. Robots here do their work in four phases. In the first phase, robots identify their axis of symmetry by identifying suitable minimal gap between two successive robots present in the ring. In this paper, it is assumed that only one such suitable gap is present; they create exactly two multiplicity points at symmetrical positions on the ring with respect to its axis of symmetry. In the second phase, robots decide their North Pole on that arc between those two multiplicity points (at symmetrical positions) which contain larger number of robots. If both the arcs between these two multiplicity points contain equal number of robots, then tiebreaking policy is described in the paper. Two nearest robots on both sides to the South Pole (opposite to the detected North Pole) are identified as guards. Except guards, all the robots join their respective multiplicity point pairwise. The axis of symmetry is now redefined by looking at the multiplicity points. In the third phase, all the robots present at the multiplicity points gather on the North Pole. Once again the axis of symmetry is determined in this phase with respect to the positions of guards. In the fourth phase, guards join the only multiplicity point at North Pole.

D'Angelo et al. [4] addressed the problem of gathering of six robots on a ring placed symmetrically on a ring with respect to their axis of symmetry (only one). This symmetry can be of *node–node* symmetry or *node–edge* symmetry but not the edge–edge symmetry. D'Angelo et al. [5] extended the solution by addressing the gathering on a ring starting from *node–node* or *node–edge* symmetry $3 \leq k < (n - 4)$, where n is number of nodes and k is number of robots ($k \neq 4$). Klasing et al. [13] mentioned that gathering of two robots is not possible on a ring. They have also mentioned that gathering in ring for robots between 4 and 18 in number needs additional characterization.

3   Conclusion and Future Scope

Research on swarm robotics in the discrete domain is in a nascent stage. Though the gathering problem is extensively studied in continuous domain, there is a considerable research gap in the discrete domain. Taking into consideration variations in graph topology such as tree, grid, bipartite graph, ring including the solvability issues of gathering problem under various robot capabilities, an attempt has been made in this paper to critically review the discrete domain gathering problem vis-à-vis the current status. Moreover, some of the impossibility results also are listed on the basis of periodicity, symmetry, etc. However, the solvability of the gathering problem for other graph topologies or for general graphs is yet to be thoroughly investigated.

References

1. Absho, S., Cord-Landwehr, A., Fischer, M., Jung, D., Meyer auf der Heide, F.: Gathering a closed chain of robots on a grid. In: IEEE International Parallel and Distributed Processing Symposium (IPDPS), pp. 689–699 (2016)
2. Bhaumik, S., Chaudhuri, S.G.: Gathering of asynchronous mobile robots in a tree. In: IEEE 2nd International Conference, Applications and Innovations in Mobile Computing (AIMoC), pp. 97–102 (2015)
3. Cord-Landwehr, A., Fischer, M., Jung, D., Meyer auf der Heide, F.: Asymptotically optimal gathering on a grid. In: 28th ACM Symposium on Parallelism in Algorithms and Architectures (SPAA '16), pp. 301–312, ACM, New York (2016)
4. D'Angelo, G., Di Stefano, G. D., Navarra, A.: Gathering of six robots on anonymous symmetric rings. In: Kosowski, A., et al. (eds.) Structural Information and Communication Complexity, SIROCCO 2011, LNCS, vol. 6796, pp. 174–185. Springer, Berlin (2011)
5. D'Angelo, G., Navarra, A., Nisse, N.: Gathering and exclusive searching on rings under minimal assumptions. In: Chatterjee, M., et al. (eds.) Distributed Computing and Networking, ICDCN 2014, LNCS, vol. 8314, pp. 149–164. Springer, Berlin (2014)
6. D'Angelo, G., Di Stefano, G., Klasing, R., Navarra, A.: Gathering of robots on anonymous grids and trees without multiplicity detection. Theoret. Comput. Sci. **610**, 158–168 (2016)
7. Di Stefano, G., Navarra, A.: Optimal gathering on in nite grids. In: Felber, P., et al. (eds.) Stabilization, Safety, and Security of Distributed Systems, SSS 2014. LNCS, vol. 8756, pp. 211–225. Springer, Berlin (2014)
8. Dutta, D., Dey, T., Chaudhuri, S.G.: Gathering multiple robots in a ring and an in nite grid. In: Krishnan, P., et al. (eds.) Distributed Computing and Internet Technology, ICDCIT 2017. LNCS, vol. 10109, pp. 15–26. Springer, Cham (2017)
9. Fischer, M., Jung, D., Meyer auf der Heide, F.: Gathering anonymous, oblivious robots on a grid. In: Fernandez Anta, A. (eds.) Algorithms for sensor systems, ALGOSENSORS 2017. LNCS, vol. 10718, pp. 168–181. Springer, Cham (2017)
10. Fraigniaud, P., Pelc, A.: Deterministic rendezvous in trees with little memory. In: Taubenfeld, G. (ed.) Distributed Computing, DISC 2008. LNCS, vol. 5218, pp. 242–256. Springer, Berlin (2008)
11. Guilbault, S., Pelc, A.: Gathering asynchronous oblivious agents with local vision in regular bipartite graphs. In: Kosowski, A., et al. (eds.) Structural Information and Communication Complexity, SIROCCO 2011. LNCS, vol. 6796, pp. 162–173. Springer, Berlin, Heidelberg (2011)

12. Klasing, R., Markou, E., Pelc, A.: Gathering Asynchronous Oblivious Mobile Robots in a Ring. In: Asano, T. (ed.) Algorithms and Computation, ISAAC 2006, LNCS, vol. 4288, pp. 744–753. Springer, Berlin, Heidelberg (2006)
13. Klasing, R., Kosowski, A., Navarra, A.: Taking advantage of symmetries: gathering of asynchronous oblivious robots on a ring. In: Baker, T.P., et al. (eds.) Principles Of Distributed Systems, OPODIS 2008. LNCS, vol. 5401, pp. 446–462. Springer, Berlin (2008)
14. Rajsbaum, S., Castaneda, A., Penaloza, D.F., Alcantara, M.: Fault-tolerant robot gathering problems on graphs with arbitrary appearing times. In: IEEE International Parallel and Distributed Processing Symposium (IPDPS), pp. 493–502 (2017)
15. Shibata, M., Ooshita, F., Kakugawa, H., Masuzawa, T.: Move-optimal partial gathering of mobile agents in asynchronous trees. Theoret. Comput. Sci. **705**, 9–30 (2018)
16. Sinha, M., Mukhopadhyaya, S.: Optimal tree search by a swarm of mobile robots. In: Mishra, D., et al. (eds.) Information and Communication Technology. Advances in Intelligent Systems and Computing, vol. 625, pp. 179–187. Springer, Singapore (2018)

Modified SA Algorithm for Bi-objective Robust Stochastic Cellular Facility Layout in Cellular Manufacturing Systems

Ravi Kumar and Surya Prakash Singh

Abstract In this era of globalization, product demand and product mix disparate frequently. In addition, entry of new product and deletion of existing product due to changing market scenarios make the manufacturing environment uncertain. In the uncertain manufacturing environment, a facility layout design must be capable to handle all these changes while keeping minimal material handling cost (MHC) and re-configuration cost. This paper proposed a novel modified simulated annealing (modified SA) approach to solving bi-objective robust stochastic cellular facility layout problem (RSCFLP). The RSCFLP minimizes material handling distance (MHD) and thus MHC and maximizes similarity score for multi-periods and provides a robust layout design considering stochastic demand for multi-periods. The proposed robust layout design suits for all time periods and avoids re-configuration cost. The proposed modified SA is tested using twenty-five data sets with a varying number of machines, products, cells, time periods, and product demand. Lastly, results are compared to show the significance of proposed approach.

Keywords Facility layout · Robust layout · Cellular manufacturing systems Simulated annealing

1 Introduction

Facility layout design problem assigns a number of machines to a number of locations in a way that overall movement of both men and materials among machines can be reduced. Ariafar and Ismail [1]; Ripon et al. [2]; and Kumar and Singh [3] highlighted that some of the major objectives linked while designing facility layout are MHC, similarity score, and noise disturbance. Further, the

R. Kumar (✉) · S. P. Singh
Department of Management Studies, Indian Institute of Technology Delhi, New Delhi, India
e-mail: ravikumar.iitdu@gmail.com

S. P. Singh
e-mail: surya.singh@gmail.com

© Springer Nature Singapore Pte Ltd. 2019   19
J. K. Mandal et al. (eds.), *Advanced Computing and Communication Technologies*,
Advances in Intelligent Systems and Computing 702,
https://doi.org/10.1007/978-981-13-0680-8_3

classification of facility layout problems is given as static facility layout problem (SFLP), dynamic facility layout problem (DFLP), and robust facility layout problem (RFLP) [4]. Also, there are subclassifications as single-row facility layout problem (SRFLP) and multi-row facility layout problem (MRFLP) [5]. In this work, a RFLP is designed for cellular manufacturing systems (CMS). Cellular manufacturing (CM) implements group technology (GT), a manufacturing philosophy. To design layout in CMS, three phases of GT are adopted. The first phase is to make part families and machine families, the second phase is to design inter-cell layout, and the final phase is to design intra-cell layout. Cellular layout design has many advantages over traditional product and process layout such as reduced movement of men and materials, reduction in lead time, and reduction in setup time. Also, cellular layout improves the productivity and quality [6, 7]. For detailed study on cellular layouts problem, readers can refer [8, 9].

Since in today's competitive environment, product demand is stochastic in nature, which forced a layout design to be robust. A robust layout design avoids re-arrangement cost and minimizes expected demand for multi-time periods [10]. Also, Singh and Singh [11] advocated that change in existing layout incurred a huge cost in long run. Moreover, layout design is a non-polynomial hard (NP-hard) problem, which cannot be solved in real time using exact approach for larger size problem. To solve larger size facility layout problem (FLP), sub-optimal approaches, i.e., heuristics and meta-heuristics approaches, are used. The paper proposes a modified SA to solve bi-objective mathematical model to design RSCFLP. The two objectives considered in RSCFLP are minimization of MHC and maximization of similarity score. The stochastic demand of products over planning horizon is considered as normally distributed. The bi-objective RSCFLP is solved using exact and modified simulated annealing (modified SA)-based approach for first twenty-one data sets, and further four larger data size problems are solved using only modified SA approach to minimize MHC and maximize similarity score for multi-time period.

The remaining of the paper is organized as follows: Section 2 provides the literature review. Section 3 explains proposed modified SA algorithm, and numerical illustration is discussed in Sect. 4. Results and discussions are provided in Sect. 5, while limitations and future scope are given in Sect. 6. Finally, Sect. 7 provides research implications.

2 Background

2.1 Literature Review of Robust Facility Layout

Rosenblatt and Lee [12] invented the term robustness to answer the uncertainty linked with facility layout problem. They considered product demand as stochastic random variable. Also, they highlighted that a robust layout may not be optimal but reliable in all the time periods for different production scenarios. They mathematically modeled the SRFLP as quadratic assignment problem (QAP). Kouvelis and Kiran [13] designed layout as QAP and dynamic programming (DP) for single and multiple periods in stochastic scenario, respectively. They assigned probabilities to different production scenarios to find an efficient alternative layout design. Kouvelis et al. [14] advocated the effect of product mix on layout design. They also proposed a QAP model to design robust layout considering demand uncertainty. Cheng et al. [15] used genetic algorithm (GA) to solve layout under uncertain environment. They consider material flow among departments as fuzzy number. Smith and Norman [16] solved block layout considering statistical percentage of MHC which is a function of mean and variance of demand. They solved the proposed model using GA. Aiello and Enea [17] designed robust layout considering production uncertainties to minimize the fuzzy MHC. Braglia et al. [18] designed single-row layout considering uncertain production rate. They handled uncertainty in production rate considering normally distributed demand. Kulturel-Konak et al. [19] desegregated robustness in layout design considering product demand independent of any specific distribution. Braglia et al. [20] also designed robust flexible layout considering stochastic demand over the planning horizon. They handled uncertainty by taking average and variance of demand for multi-periods. Tavakkoli-Moghaddam [21] designed cellular layout problem solving proposed mathematical model considering stochastic demand. Decision maker's attitude was also considered while solving the proposed mathematical model. Krishnan et al. [22] proposed a robust layout for multi-time period to minimize the maximum loss in MHC. The proposed model was solved using GA. Later, the proposed model was extended to minimize the total expected loss. Jithavech and Krishnan [23] predicted uncertainties while designing layout using simulation approach. Pillai et al. [24] designed RSCFLP using SA to minimize total penalty cost (TPC). Ariafar et al. [25] proposed a mathematical model to design stochastic facility layout in CMS considering uniform distribution of material flow. They solved the proposed model on LINGO considering different cumulative demand probability. Ariafar et al. [26] designed stochastic facility layout considering product demand uncertainties. The proposed mathematical model minimized both inter-/intra-cell MHC. Zhao and Wallace [27] developed a greedy flow-map heuristic to solve stochastic facility layout considering uncertain demand level. They claimed that that the proposed approach provides randomness at marginal increase in computational costs. Neghabi et al. [10] incorporated uncertainty considering unknown dimensions of departments to design robust facility layout. They also designed an adaptive algorithm to incorporated decision makers' requirements. Li et al. [28] proposed a modified SA approach to reduce intercellular MHD considering uncertainty of remanufacturing due to stochastic return of used product or components. Sakhaii et al. [29] formulated layout problem as mixed-integer linear program (MILP) considering part processing time uncertainty and unreliable machines. The proposed model was solved using CPLEX. Zha et al. [30] designed robust layout considering uncertain product demands to minimize MHC. They considered unequal area department size. Also, Balakrishnan and Cheng [31], Moslemipour et al. [4] can be referred for uncertainty in RFLP.

2.2 Research Gaps

As discussed in the earlier section, the cellular manufacturing has many benefits linked over traditional product and process layouts. However, in the literature, there is limited work on CMS considering uncertain product demand. Also, most of the work in CMS is focused toward single objective. Moreover, most of the work in CMS is proposed using smaller size problems. This paper proposed a novel modified SA approach to solve a bi-objective RSCFLP. The proposed modified SA is capable of solving larger size problems with different combinations of products, machines, cells, and time.

3 SA-Based Meta-heuristic

3.1 Problem Statement

Kumar and Singh [32] proposed a bi-objective robust stochastic model to design robust cellular facility layout. However, due to NP-hard nature of the problem, the proposed model was complex to solve optimally in polynomial time for larger size problems. The proposed mathematical model is given below with all notations and assumptions.

List of Indices

i, k	Refer to machines
j, l	Refer to locations
p, q	Refer to cells
t	Refer to the time periods
n	Refer to the products
r	Refer to the processes

List of Parameters

m	Number of machines or locations
c	Number of cells
N	Number of products
R	Number of processes
NMC_p	Number of machines in cell p
MF_{ik}	Stochastic flow between machine i and machine k
Min Z	Objective function value
f_{ikt}	Flow between machine i and machine k for time period t

$\mathrm{Exp}(f_{ikt})$ Expected value of flow over planning horizon (i.e., for multi-period flow)

$\mathrm{Var}(f_{ikt})$ Variance of flow over the planning horizon

Zp Standard normal z values for percentile p

d_{jl} Distance between location j and location l

D_{nt} Demand of product n in time period t

A_{ik} Stochastic similarity score between machine i and machine k

S_{ikt} Similarity score between machine i and machine k for time period t

$\mathrm{Exp}(S_{ikt})$ Expected value of similarity score over the planning horizon (i.e., for multi-period similarity score

$\mathrm{Var}(S_{ikt})$ Variance of similarity score over the planning horizon

F_{nrit} $\begin{cases} 1 & \text{if product } n \text{ in process } r \text{ on machine } i \text{ in time period } t \\ 0 & \text{otherwise} \end{cases}$

a_{ikt} $\begin{cases} 1 & \text{if product } n \text{ is being processed on both machines } i \text{ and } k \text{ in time } t \\ 0 & \text{otherwise} \end{cases}$

a_{iit} $\begin{cases} 1 & \text{if product } n \text{ is being processed only on machine } i \text{ in time } t \\ 0 & \text{otherwise} \end{cases}$

a_{kkt} $\begin{cases} 1 & \text{if product } n \text{ is being processed only on machine } k \text{ in time } t \\ 0 & \text{otherwise} \end{cases}$

List of Variables

$$X_{ij} \begin{cases} 1 & \text{if machine } i \text{ is located at location } j \\ 0 & \text{otherwise} \end{cases}$$

$$Y_{ip} \begin{cases} 1 & \text{if machine } i \text{ is located in cell } p \\ 0 & \text{otherwise} \end{cases}$$

List of Assumptions

- Machines are considered of equal size.
- Stochastic demand is considered as normally distributed over the planning horizon.
- Cost per part per distance is considered as one.
- Distance between each pair of location is known.

Bi-objective RSCFLP

$$\operatorname{Min} Z = \sum_{i,k}^{m} \sum_{j,l}^{m} \sum_{p,q}^{c} \mathrm{MF}_{ik} d_{jl} X_{ij} X_{kl} Y_{ip} Y_{kq} - \sum_{i,k}^{m} \sum_{\substack{p,q \\ p \neq q}}^{c} A_{ik} Y_{ip} Y_{kq} \tag{1}$$

$$\mathrm{MF}_{ik} = \mathrm{Exp}(f_{ikt}) + Zp * \sqrt{\mathrm{Var}(f_{ikt})^2} \tag{1a}$$

$$f_{ikt} = \sum_{n,r}^{N,R} D_{nt} * F_{nrit} * F_{n(r+1)kt} \tag{1b}$$

$$A_{ik} = \mathrm{Exp}(S_{ikt}) + Zp * \sqrt{\mathrm{Var}(S_{ikt})^2} \tag{1c}$$

$$S_{ikt} = \frac{\sum_{n=1}^{N} a_{ikt}}{\sum_{n=1}^{N} (a_{ikt} + a_{iit} + a_{kkt})} \tag{1d}$$

Subject to,

$$\sum_{i=1}^{m} X_{ij} = 1 \qquad \forall j \tag{2}$$

$$\sum_{j=1}^{m} X_{ij} = 1 \qquad \forall i \tag{3}$$

$$\sum_{p=1}^{c} Y_{ip} = 1 \qquad \forall i \tag{4}$$

$$\sum_{p=1}^{c} Y_{ip} = \mathrm{NMC}_p \qquad \forall p \tag{5}$$

Equation (1) minimizes the MHC and maximizes the similarity score. The first part of Eq. (1) shows the MHC, and the second part shows similarity score. Aggregated stochastic material flow among machines for all time periods is calculated using Eq. (1a). The aggregated stochastic material flow is a function of expected material flow and variance of material flow among machines. Equation (1b) calculates flow among machines for each time period using demand and part processing data for each time period. Equation (1c) calculates aggregated stochastic similarity score among machines for all the time periods. The aggregated similarity score is a function of expected similarity score and variance of similarity score among machines. Equation (1d) calculates similarity score between each pair of machines for every time period using machine part data. Equation (2) ensures that a single machine assigns to a single location, and Eq. (3) ensures that each location must only assign to a single machine. Equation (4) constrains machines to overlap among cells. Equation (5) bounds the number of machines within each cell.

3.2 Modified SA

It is hard to solve the FLP using exact approaches in polynomial time due to its complex nature. To solve larger size facility layout design problems, sub-optimal approaches such as heuristics and meta-heuristics are developed. In this work, to solve the bi-objective RSCFLP discussed in the previous section, a modified SA is developed and coded using MATLAB 2014 software. SA is a stochastic search procedure to search for good sub-optimal solution. To start SA algorithm, we need an initial solution, a random generate of moves, an annealing schedule, a quantitative objective function, and a termination criteria [33]. The initial solution for bi-objective RSCFLP is randomly generated, i.e., machine-location assignment and machine-cell assignment. Then, the value of quantitative objective function (Min Z) represented in Eq. (1) of Sect. 3.1 is computed for the randomly generated initial solution. Then with a random move, nearby solution is generated and again the objective function value is computed (Min Z'). This random move to generate neighborhood solution is done by swapping machine locations. Now, the difference of both objective function values, i.e., $\Delta Z = $ Min $Z' - $ Min Z, is computed. If the ΔZ value is negative, then the new solution is accepted since the problem is of minimization; otherwise, solution is accepted with an acceptance probability. This process is iterated till the termination criteria are met. This SA approach is embedded with a self-iterative process to design modified SA. This embedded approach runs the SA algorithm for N number of trials, and for each trial, it stores the objective function value and corresponding solution. It compares the objective function value of current trial (OV_New) with the objective function value of the previous trial (OV_Best). It updates the current objective function value (OV_New) with the previous objective function value (OV_Best) if the objective function value of the previous trial was less in case of minimization, and also updates the solution of the previous trial. So, after the N trials are over, it gives the minimum objective function value and corresponding solution out of N trials of SA. The pseudocode for the proposed modified SA can be seen in Fig. 1. For detailed study on SA, papers by Kirkpatrick et al. [33]; Van Laarhoven and Aarts [34]; and Tayal and Singh [35] can be referred.

For N_trial = 1: N
>   [New_Solution, OV_New]=simulated_mincon
Check
>   If (OV_New < OV_Best)
Update
>   Best_Solution = New_Solution
>   OV_Best = OV_New
End
}
>   Display
>   OV_Best
>   Best_Solution

}
End
>   simulated_mincon
Initialise
>   A known or randomly generated initial solution, $v0$
>   Set up the initial temperature, $T\_initial$
>   Set up the final stopping temperature, T_min
>   Set up maximum number of rejection, Max_rej
>   Set up maximum number of runs, Max_run
>   Set up maximum number of accept, Max_accept
>   Set up initial search period, intial_search
Generate neighborhood solution, ns
>   Find out a new feasible solution by randomly swapping two machines and their positions,
>   ns = new randomly generated solution
Start inner loop
>   Find out the OV for both initially generated solution (v0) and neighborhood solution (ns) using
mathematical model provided in section 3.1.4.
Check
>   If $-(Min Z'(ns) - Min Z(v0)) > E\_norm$,
Update
>   v0=ns
>   Else if $(Min Z'(ns) - Min Z(v0)) < E\_norm$, &&
> $\exp[-(Min Z'(ns) - Min Z(v0))/KT)] > rand$,
>   rand ? (0,1)
Repeat
>   Until inner loop criteria
Decrease the temperature, using cooling schedule function
Reset inner loop criteria
Repeat
>   Until stopping criteria (T < T_min) or (inner loop >= Max_rej)
Output
>   the best solution (layout) and corresponding objective function value
End

Fig. 1 Modified SA pseudocode to solve bi-objective RSCFLP

Initial Temperature

To keep the probability of initial acceptance of moves near to 1, the initial temperature must be large [33]. In this work, initial temperature (T_initial) is calculated by Eq. (6).

$$T\_initial = \frac{(Z(\min) - Z(\max))}{\ln(P)} \tag{6}$$

$Z(\min)$ and $Z(\max)$ values for each data set are predetermined using linear cooling schedule. The number of trials considered to determine these values is taken as hundred. The $Z(\min)$ and $Z(\max)$ values to calculate T_initial are provided as supplementary file.

Cooling function

Lundy and Mees [36] provided a cooling function and advocated that this cooling function results in slower cooling than other cooling functions. This work computes cooling function value using Eq. (7) given by Lundy and Mees [36]. The β value is calculated using Eq. (8) given by Connolly [37]. M denotes trials for each temperature level.

$$T_k = \frac{T_{k-1}}{(1 + \beta T_{k-1})} \tag{7}$$

$$\beta = \frac{(T\_initial - T\_min)}{\{(M-1) * T\_initial * T\_min\}} \tag{8}$$

Trials at each temperature level

Equation (9) calculates the number of trials for each temperature level which should be close to problem size as suggested by Park and Kim [38].

$$\text{Max\_run (trials)} = M * T * C \tag{9}$$

Stopping criteria

Van Laarhoven et al. [39] suggested that when temperature arrives at a predetermined temperature, SA algorithm can be stopped. It can also be stopped when it arrives at predetermined number of trials as suggested by Park and Kim [38]. The proposed modified SA can be terminated when it reaches either of the above-mentioned conditions.

4   Numerical Illustrations

To approve and check complex nature of proposed meta-heuristic, twenty-one data sets from Kumar and Singh [3] are considered. These data sets consist of different combinations of products, machines, cells, and time. These data sets are solved utilizing exact and proposed meta-heuristic approach. Moreover, four large data sets are solved only using proposed meta-heuristic since exact method is unable to handle these large data sets. Further, four larger data sets viz. Data Set 22, Data Set 23, Data Set 24, and Data Set 25 are generated randomly and can be provided by authors on request. The notation of these four large data sets is as Data Set 22 ($T = 3$, $P = 7$, $M = 15$), Data Set 23 ($T = 3$, $P = 7$, $M = 20$), Data Set 24 ($T = 3$, $P = 7$, $M = 25$), Data Set 25 ($T = 3$, $P = 7$, $M = 30$), respectively. The modified SA approach is run for 1000 trials; i.e., N is taken 1000 to run the simulation. Objective function value and CPU time of all twenty-five data sets are tabulated for exact and meta-heuristic. Both exact and modified SA approach run on Core i5 Processor with 4 GB RAM in Windows environment. Exact approach simulations are run on LINGO 10, and modified SA simulations are run on MATLAB 2014.

5   Results and Discussion

Since FLP is combinatorial optimization, so it is hard to solve in reasonable time. Due to its NP-hard nature, it requires sub-optimal approaches to be developed. This paper proposed a modified SA approach to solve the bi-objective RSCFLP optimally in polynomial time. The RSCFLP to design robust stochastic cellular facility layout consists of two major quantitative objectives of facility layout design problem. The first objective is to minimize the MHD for multi-time period to reduce the MHC. The second objective is to maximize the overall similarity score to group the machines into cells. So the proposed model groups machines into cells and minimizes inter-/intra-cell movement. Then, to validate and check the complexity of proposed modified SA, Data Set 1 to Data Set 21 are solved using both exact and modified SA, and it is found that for all twenty-one data sets, modified SA provides the optimal solution in reasonable computational time as compared to exact approach. For larger size problems, i.e., Data Set 22, Data Set 23, Data Set 24, and Data Set 25, exact approach is unable to provide the solution in reasonable time due high complexity of the problem; however, modified SA provides solution for larger size problems, i.e., Data Set 22, Data Set 23, Data Set 24, and Data Set 25 in reasonable time. The detailed solutions of Data Set 1 to Data Set 25 are provided in Table 1. Finally, objective function value and CPU time of all twenty-five data sets for exact and modified SA are provided in Table 2. The detailed solution of each data set is provided as supplementary material.

Table 1 Solution obtained from Data Set 1 to Data Set 25

Machine-location and machine-cell assignment of each time period for all data sets, i.e., X_{ij} and $Y_{t1}, Y_{t2}, ..., Y_{tp}$

Data Set 1 ($T = 2, P = 5,$ $M = 5, C = 2$)	Data Set 2 ($T = 3, P = 5,$ $M = 5, C = 2$)	Data Set 3 ($T = 4, P = 5,$ $M = 5, C = 2$)
Machine assignment = {5, 3, 1, 4, 2}[a] Cell assignment = {(3, 4, 5), (1, 2)}[b]	Machine alignment = {5, 3, 2, 4, 1} Cell assignment = {(3, 4, 5), (1, 2)}	Machine alignment = {5, 4, 1, 3, 2} Cell assignment = {(1, 2, 5), (3, 4)}
Data Set 4 ($T = 2, P = 5,$ $M = 5, C = 2$)	Data Set 5 ($T = 3, P = 5,$ $M = 5, C = 2$)	Data Set 6 ($T = 4, P = 5,$ $M = 5, C = 2$)
Machine alignment = {1, 3, 5, 2, 4} Cell assignment = {(3, 4, 5), (1, 2)}	Machine alignment = {5, 4, 2, 3, 1} Cell assignment = {(3, 4, 5), (1, 2)}	Machine alignment = {3, 1, 2, 5, 4} Cell assignment = {(1, 2, 5), (3, 4)}
Data Set 7 ($T = 2, P = 5,$ $M = 5, C = 2$)	Data Set 8 ($T = 3, P = 5,$ $M = 5, C = 2$)	Data Set 9 ($T = 4, P = 5,$ $M = 5, C = 2$)
Machine alignment = {4, 5, 3, 1, 2} Cell assignment = {(3, 4, 5), (1, 2)}	Machine alignment = {3, 5, 4, 2, 1} Cell assignment = {(3, 4, 5), (1, 2)}	Machine alignment = {2, 1, 3, 4, 5} Cell assignment = {(1, 2, 5), (3, 4)}
Data Set 10 ($T = 2, P = 5,$ $M = 7, C = 2$)	Data Set 11 ($T = 2, P = 5,$ $M = 7, C = 3$)	Data Set 12 ($T = 2, P = 5,$ $M = 7, C = 2$)
Machine alignment = {1, 2, 4, 3, 5, 6, 7} Cell assignment = {(3, 4, 7), (1, 2, 5, 6)}	Machine alignment = {1, 2, 4, 3, 5, 6, 7} Cell assignment = {(3, 4), (1, 2), (5, 6, 7)}	Machine alignment = {1, 2, 4, 3, 5, 6, 7} Cell assignment = {(3, 4, 7), (1, 2, 5, 6)}
Data Set 13 ($T = 2, P = 5,$ $M = 7, C = 3$)	Data Set 14 ($T = 2, P = 5,$ $M = 7, C = 3$)	Data Set 15 ($T = 2, P = 5,$ $M = 7, C = 3$)
Machine alignment = {1, 3, 5, 2, 4, 7, 6} Cell assignment = {(3, 4), (1, 6), (2, 5, 7)}	Machine alignment = {1, 2, 4, 3, 5, 6, 7} Cell assignment = {(3, 4), (1, 2), (5, 6, 7)}	Machine alignment = {1, 2, 4, 3, 5, 6, 7} Cell assignment = {(3, 4), (1, 6), (2, 5, 7)}
Data Set 16 ($T = 2, P = 5,$ $M = 8, C = 2$)	Data Set 17 ($T = 2, P = 5,$ $M = 8, C = 2$)	Data Set 18 ($T = 2, P = 5,$ $M = 8, C = 2$)
Machine alignment = {5, 3, 2, 7, 6, 1, 4, 8} Cell assignment = {(1, 2, 5), (3, 4, 6, 7, 8)}	Machine alignment = {5, 3, 2, 7, 6, 1, 4, 8} Cell assignment = {(1, 2, 5), (3, 4, 6, 7, 8)}	Machine alignment = {5, 3, 2, 7, 6, 1, 4, 8} Cell assignment = {(1, 2, 5), (3, 4, 6, 7, 8)}
Data Set 19 ($T = 2, P = 7,$ $M = 8, C = 2$)	Data Set 20 ($T = 2, P = 7,$ $M = 5, C = 2$)	Data Set 21 ($T = 2, P = 7,$ $M = 8, C = 2$)
Machine alignment = {5, 3, 4, 7, 6, 2, 1, 8} Cell assignment = {(4, 5, 7), (1, 2, 3, 6, 8)}	Machine alignment = {4, 6, 7, 2, 3, 8, 5, 1} Cell assignment = {(2, 3, 8), (1, 4, 5, 6, 7)}	Machine alignment = {4, 6, 7, 2, 3, 8, 5, 1} Cell assignment = {(5, 7, 8), (1, 2, 3, 4, 6)}

(continued)

Table 1 (continued)

Machine-location and machine-cell assignment of each time period for all data sets, i.e., X_{ij} and $Y_{t1}, Y_{t2}, \ldots, Y_{tp}$

Data Set 22 ($T = 3, P = 7, M = 15, C = 3$)

Machine assignment = {1, 6, 7, 10, 15, 3, 5, 8, 11, 14, 2, 4, 9, 12, 13}
Cell assignment = {(1, 3, 5, 7, 12, 14, 15), (2, 8, 9, 10, 11), (4, 6, 13)}

Data Set 23 ($T = 3, P = 7, M = 20, C = 4$)

Machine assignment = {5, 17, 7, 16, 20, 13, 10, 9, 6, 12, 19, 11, 18, 14, 15, 8, 1, 3, 2, 4}
Cell assignment = {(2, 7, 8, 12, 14, 15, 19), (4, 5, 10, 17, 18), (6, 13, 16, 20), (1, 3, 9, 11)}

Data Set 24 ($T = 3, P = 7, M = 25, C = 5$)

Machine assignment = {23, 22, 24, 25, 17, 2, 21, 19, 18, 15, 3, 5, 20, 16, 14, 1, 6, 8, 11, 13, 4, 7, 9, 10, 12}
Cell assignment = {(1, 2, 10, 12, 17, 18, 19, 25), (4, 7, 8, 14, 15, 20, 23), (6, 13, 16, 22, 24), (3, 5, 9, 11, 21)}

Data Set 25 ($T = 3, P = 7, M = 30, C = 5$)

Machine assignment = {8, 9, 10, 11, 13, 19, 20, 25, 30, 29, 6, 4, 7, 12, 14, 16, 18, 22, 27, 28, 5, 3, 2, 1, 15, 17, 21, 23, 24, 26}
Cell assignment = {(4, 5, 6, 9, 12, 15, 16, 26), (1, 10, 17, 19, 24, 25, 27), (42, 13, 22, 28, 29), (7, 8, 14, 21, 23), (3, 11, 18, 20, 30)}

{5, 3, 1, 4, 2} [a]Represents machine assignment in locations 1, 2, 3, 4, 5, respectively
[b]Represents machines 3, 4, 5 are assigned in cell 1 and machines 1, 2 are assigned in cell 2

Table 2 OV and CPU time for exact and modified SA	Data sets	Exact approach		Modified SA	
		OV	CPU time (hh:mm:ss)	OV	Time (s)
	1	1122	1:43	1122	39.539
	2	1174	0:46	1174	4.750
	3	1184	1:52	1184	3.208
	4	2188	1:37	2188	1.858
	5	2734	2:00	2734	2.648
	6	3014	48 s	3014	3.678
	7	3703	51 s	3703	1.694
	8	4668	1:07	4668	2.685
	9	5319	0:47	5319	3.144
	10	2738	57:08:00	2738	139.074
	11	2734	52:12:27	2734	221.231
	12	5014	53:48:00	5014	170.866
	13	5011	44:27:32	5011	235.526
	14	8394	3:08	8394	145.510
	15	8391	18:20:44	8391	181.928
	16	3102	37:21:45	3102	1087.775
	17	6049	94:04:30	6049	162.935

(continued)

Table 2 (continued)

Data sets	Exact approach		Modified SA	
	OV	CPU time (hh:mm:ss)	OV	Time (s)
18	10,081	89:17:32	10,081	138.595
19	4760	37:19:26	4760	163.770
20	8328	50:20:15	8328	146.617
21	15,093	58:15:17	15,093	165.569
22	[a]	[a]	6402	1115.041
23	[a]	[a]	13,775	2468.179
24	[a]	[a]	10,714	4347.412
25	[a]	[a]	13,770	8827.750

OV objective value calculated using Eq. (1), and
[a]Solver fails to solve the problem optimally

6 Limitations and Future Scope

This work proposed modified SA capable of solving larger size problems in reasonable time; however, work can further be extended in the following ways. The complexity of proposed modified SA can be checked by comparing results with different acceptance probability. In this work, 0.95 is taken as acceptance probability; other acceptance probabilities can be considered to check the complexity. Sensitivity analysis can also be done on different parameters of proposed modified SA. The proposed modified SA can also be checked for larger size and complex problems, and also it can be checked for different facility layout designs. In the proposed meta-heuristic, cooling function given by Lundy and Mees [36] is taken; however, complexity and quality of solution can also be compared considering other cooling functions. The proposed modified SA can also be extended for unequal FLP and multi-floor facility layout design.

7 Research Implications

Though the facility layout design problem is NP-hard, the proposed modified SA is capable to solve complex larger size cellular facility layout design problems in reasonable time. Since the market is competitive nowadays, product demand and product mix change frequently which lead to a dynamic behavior; however, the proposed meta-heuristic provides one-time solution which suits for all the time periods. It can also encounter all diverse mix of products, machines, cells, and time periods. Moreover, the proposed modified SA is easy to understand and implement by decision makers.

References

1. Ariafar, S., Ismail, N.: An improved algorithm for layout design in cellular manufacturing systems. J. Manuf. Syst. **28**(4), 132–139 (2009)
2. Ripon, K.S.N., Khan, K.N., Glette, K., Hovin, M., Torresen, J.: Dynamic facility layout problem under uncertainty: a Pareto-optimality based multi-objective evolutionary approach. Cent. Eur. J. Comp. Sci. **1**(4), 375–386 (2011)
3. Kumar, R., Singh, S.P.: A similarity score-based two-phase heuristic approach to solve the dynamic cellular facility layout for manufacturing systems. Eng. Optim. **49**(11), 1848–1867 (2017)
4. Moslemipour, G., Lee, T.S., Rilling, D.: A review of intelligent approaches for designing dynamic and robust layouts in flexible manufacturing systems. Int. J. Adv. Manuf. Technol. **60**(1–4), 11–27 (2012)
5. Salmani, M.H., Eshghi, K., Neghabi, H.: A bi-objective MIP model for facility layout problem in uncertain environment. Int. J. Adv. Manuf. Technol. **81**(9–12), 1563–1575 (2015)
6. Singh, N., Rajamani, D.: Cellular manufacturing systems: design, planning and control. Springer, New York (1996)
7. Wemmerlov, U., Johnson, D.J.: Cellular manufacturing at 46 user plants: implementation experiences and performance improvements. Int. J. Prod. Res. **35**, 29–49 (1997)
8. Luo, J., Tang, L.: A hybrid approach of ordinal optimization and iterated local search for manufacturing cell formation. Int. J. Adv. Manuf. Technol. **0**, 362–372 (2009)
9. Liu, C., Yin, Y., Yasuda, K., Lian, J.: A heuristic algorithm for cell formation problems with consideration of multiple production factors. Int. J. Adv. Manuf. Technol. **46**, 1201–1213 (2010)
10. Neghabi, H., Eshghi, K., Salmani, M.H.: A new model for robust facility layout problem. Inf. Sci. **278**, 498–509 (2014)
11. Singh, S.P., Singh, V.K.: An improved heuristic approach for multi-objective facility layout problem. Int. J. Prod. Res. **48**(4), 1171–1194 (2010)
12. Rosenblatt, M.J., Lee, H.L.: A robustness approach to facilities design. Int. J. Prod. Res. **25**(4), 479–486 (1987)
13. Kouvelis P, Kiran AS (1990) The plant layout problem in automated manufacturing systems. Ann. Oper. Res. **26**
14. Kouvelis, P., Kurawarwala, A.A., Gutierrez, G.J.: Algorithms for robust single and multiple period layout planning for manufacturing systems. Eur. J. Oper. Res. **63**(2), 287–303 (1992)
15. Cheng, R., Gen, M., Tozawa, T.: Genetic search for facility layout design under interflows uncertainty. In IEEE International Conference on Evolutionary Computation, 1995, vol. 1, p. 400. IEEE November 1995
16. Smith, A.E., Norman, B.A.: Evolutionary design of facilities considering production uncertainty. Evolutionary Design and Manufacture, pp. 175–186. Springer, London (2000)
17. Aiello, G.I.U.S.E.P.P.E., Enea, M.A.R.I.O.: Fuzzy approach to the robust facility layout in uncertain production environments. Int. J. Prod. Res. **39**(18), 4089–4101 (2001)
18. Braglia, M., Zanoni, S., Zavanella, L.: Layout design in dynamic environments: strategies and quantitative indices. Int. J. Prod. Res. **41**(5), 995–1016 (2003)
19. Kulturel-Konak, S., Smith, A.E., Norman, B.A.: Layout optimization considering production uncertainty and routing flexibility. Int. J. Prod. Res **42**(21), 4475–4493 (2004)
20. Braglia, M., Zanoni, S., Zavanella, L.: Layout design in dynamic environments: analytical issues. Int. Trans. Oper. Res. **12**(1), 1–19 (2005)
21. Tavakkoli-Moghaddam, R., Javadian, N., Javadi, B., Safaei, N.: Design of a facility layout problem in cellular manufacturing systems with stochastic demands. Appl. Math. Comput. **184**(2), 721–728 (2007)
22. Krishnan, K.K., Cheraghi, S.H., Nayak, C.N.: Facility layout design for multiple production scenarios in a dynamic environment. Int. J. Ind. Syst. Eng. **3**(2), 105–133 (2008)

23. Jithavech, I., Krishnan, K.K.: A simulation-based approach for risk assessment of facility layout designs under stochastic product demands. Int. J. Adv. Manuf. Technol. **49**(1–4), 27–40 (2010)
24. Pillai, V.M., Hunagund, I.B., Krishnan, K.K.: Design of robust layout for dynamic plant layout problems. Comput. Ind. Eng. **61**(3), 813–823 (2011)
25. Ariafar, S.H., Ismail, N., Tang, S.H., Ariffin, M.K.A.M., Firoozi, Z.: A stochastic facility layout model in cellular manufacturing systems. Int. J. Phys. Sci. **6**(15), 3666–3670 (2011)
26. Ariafar, S., Ismail, N., Tang, S.H., Ariffin, M.K.A.M., Firoozi, Z.: The reconfiguration issue of stochastic facility layout design in cellular manufacturing systems. Int. J. Serv. Operat. Manag. **11**(3), 255–266 (2012)
27. Zhao, Y., Wallace, S.W.: Integrated facility layout design and flow assignment problem under uncertainty. INFORMS J. Comput. **26**(4), 798–808 (2014)
28. Li, L., Li, C., Ma, H., Tang, Y.: An optimization method for the remanufacturing dynamic facility layout problem with uncertainties. In Discrete Dynamics in Nature and Society (2015)
29. Sakhaii, M., Tavakkoli-Moghaddam, R., Bagheri, M., Vatani, B.: A robust optimization approach for an integrated dynamic cellular manufacturing system and production planning with unreliable machines. Appl. Math. Model. **40**(1), 169–191 (2016)
30. Zha, S., Guo, Y., Huang, S., Wang, F., Huang, X.: Robust facility layout design under uncertain product demands. Procedia CIRP **63**, 354–359 (2017)
31. Balakrishnan, J., Cheng, C.H.: Multi-period planning and uncertainty issues in cellular manufacturing: a review and future directions. Eur. J. Oper. Res. **177**(1), 281–309 (2007)
32. Kumar, R., Singh, S.P.: Designing robust stochastic bi-objective cellular layout in manufacturing systems. Int. J. Manag. Concepts Philos. **10**(2), 147–164 (2017)
33. Kirkpatrick, S., Gelatt, C.D., Vecchi, M.P.: Optimization by simulated annealing. Science **220**(4598), 671–680 (1983)
34. Van Laarhoven, P.J., Aarts, E.H.: Simulated annealing theory and application. Kluwer Academic Publishers, USA (1987)
35. Tayal A, Singh SP (2016) Integrating big data analytic and hybrid firefly-chaotic simulated annealing approach for facility layout problem. Ann. Oper. Res. 1–26
36. Lundy, M., Mees, A.: Convergence of an annealing algorithm. Math. Program. **34**(11), 1–124 (1986)
37. Connolly, D.: An improved annealing scheme for the QAP. Eur. J. Oper. Res. **46**, 93–100 (1990)
38. Park, Moon-Won, Kim, Yeong-Dae: A systematic procedure for setting parameters in simulated annealing algorithms. Comput. Oper. Res. **25**(3), 207–217 (1998)
39. Van Laarhoven, P.J., Aarts, E.H.L., Lenstra, J.K.: Job shop scheduling by simulated annealing. Oper. Res. **40**, 113–125 (1992)

Fuzzy Time Series Forecasting Method Using Probabilistic Fuzzy Sets

Krishna Kumar Gupta and Sanjay Kumar

Abstract In the literature of time series forecasting, no method can handle both probabilistic and non-probabilistic uncertainty simultaneously. In the current investigation, we have presented probabilistic fuzzy set (PFS) based fuzzy time series (FTS) forecasting model to describe the issue of uncertainties that rises due to randomness as well as linguistic representation of time series data. An aggregation operator is also presented in this paper to aggregate the fuzzified outputs using with membership grades associated with corresponding probabilities. The presented model has been applied to forecast the time series data of University of Alabama enrolments. The performance of presented model has been examined in terms of RMSE and AFE.

Keywords Fuzzy logical relation · Uncertainties · Fuzzy time series forecasting Probabilistic fuzzy set

1 Introduction

Moving average (MA), autoregressive moving average (ARMA), regression analysis as well as exponential moving average have been utilized for time series forecasting and modelling for a long time. Major shortcomings of traditional statistical tool-based time series forecasting methods are their incapable of handling uncertainty caused by non-probabilistic reasons and linguistic representation of time series data. Time complexity and less accuracy in forecasted outputs are other issues in traditional time series forecasting method. Song and Chissom [25–27] proposed fuzzy set (FS) [33] in TS forecasting and established some forecasting

K. K. Gupta · S. Kumar (✉)
Department of Mathematics, Statistics and Computer Science, G. B. Pant,
University of Agriculture and Technology, Pantnagar 263145, India
e-mail: skruhela@hotmail.com

K. K. Gupta
e-mail: guptakrishna.gupta@gmail.com

© Springer Nature Singapore Pte Ltd. 2019   35
J. K. Mandal et al. (eds.), *Advanced Computing and Communication Technologies*,
Advances in Intelligent Systems and Computing 702,
https://doi.org/10.1007/978-981-13-0680-8_4

models to deal uncertainty in time series forecasting that arises due to vagueness, inaccuracy, and imprecision in time series data.

Chen [3] proposed simple arithmetic operators rather than complex max–min compositions operators used by Song and Chissom [25–27]. Afterwards, various fuzzy time series (FTS) forecasting methods with the innovation either in partitioning the domain of discourse or in fuzzy logical relations (FLRs) to improve the correctness in forecast are proposed by Chen and Tanuwijaya [5], Chen and Hwang [4], Huarng [16], Song [28], Lee and Chou [20], Liu [23], Cheng et al. [10, 11]. Many researchers Huarng and Yu [17], Chen and Kao [6], Chen and Chen [7], Yolcu [32], Ye et al. [30], Deng et al. [12], Chen and Phuong [8], Xian et al. [29] proposed various FTS forecasting models using swarm optimization, support vector machine, granular computing, and other machine learning techniques to develop adaptive and intelligent FTS forecasting models.

Probabilistic and non-probabilistic uncertainties are two conceptually different kinds of uncertainties which occur simultaneously in the system. The main advantage of FTS forecasting models is their ability for handling non-stochastic uncertainty. However, these forecasting models do not hold the capabilities to manage the stochastic uncertainties. Meghdadi [24] introduced probabilistic fuzzy set (PFS) to consider both uncertainties in a single framework. Its main advantage of combining interpretability of fuzzy set with statistical properties for control problems and modelling in a probabilistic fuzzy logic system was proposed by Liu and Li [22]. Applications of PFS were explored by many researchers Hinojosa et al. [15], Li and Huang [21], Almeida et al. [1], Fialho et al. [14] in various fields where probabilistic uncertainty plays an equal important role as non-probabilistic uncertainty.

In this study, we present a FTS forecasting model that includes both probabilistic and non-probabilistic uncertainties in a single framework using PFS. PFS is constructed using a probability distribution function that associates probabilities to possible membership grades of time series data in FSs. We introduce an aggregation operator to aggregate the fuzzified outputs using corresponding probabilities. In order to verify the performance of the proposed forecasting model, it has been implemented to forecast the University of Alabama enrolments.

2   Preliminaries

FS, PFS, and FTS are defined in this section as follows:

Definition 1 [33] Let $X = \{x_1, x_2, \ldots, x_n\}$ be a finite domain of discourse. A fuzzy set F on $X = \{x_1, x_2, \ldots, x_n\}$ is defined as follows:

$$F = \{\langle x, \mu_F(x)\rangle | \forall x \in X\} \tag{1}$$

where $\mu_F : X \to [0, 1]$ and $\mu_F(x)$ represent membership grade of x in F.

Definition 2 [22] For variable $x \in X$ with membership grade $(0 \leq \mu \leq 1)$, PFS F' may be described by probability space $(V_x, \Im, P)$, where V_x is collection of every eventual event $(0 \leq \mu \leq 1)$ and probability P is assigned on $\Im(\sigma - \text{field})$. All B_i's are in V_x

$$P(\text{B}_i) \geq 0, \quad P\left(\sum \text{B}_i\right) = \sum P(\text{B}_i), \quad P(V_x) = 1 \tag{2}$$

where $P(B_i)$ is the probability of the event B_i and B_i is a membership grade $\mu = \mu_i(i = 1, 2, \ldots, n)$ is subset of $[0, 1]$, and n is the index of element event in the set $(0 \leq \mu \leq 1)$. The PFS F' may be written in terms of the union of finite sub-probability space defined as:

$$F' \equiv \bigcup_{x \in X} (V_x, \Im, P). \tag{3}$$

Definition 3 [3] Let $Z(t)(t \in \mathbb{Z})$ be subset of $\mathbb{R}$ which is the domain of discourse on which FSs $h_i(t)(i \in \mathbb{N})$. A FTS $H(t)$ on $Z(t)$ is a set of FS $h_i(t)$. If $\exists H(t) = H(t-1) \circ R(t-1, t)$, here, "$\circ$" is max–min composition operator, $R(t-1, t)$, is a fuzzy logical relationship (FLR), FLR $H(t-1) \rightarrow H(t)$ between $H(t)$ and $H(t-1)$ represents that $H(t)$ is caused by $H(t-1)$. If $H(t)$ is caused by more FSs $H(t-1), H(t-2), \ldots, H(t-m)$, then this FLR is a mth-order FTS $H(t-m), \ldots, H(t-2), H(t-1) \rightarrow H(t)$.

3 Proposed Method and Experimental Study

This section explains the process of proposed probabilistic FTS forecasting method and experimental result of its implementation to forecast University of Alabama enrolments. Its stepwise procedure is described as follows:

Step 1: Let $X = [D_{\min} - \sigma, D_{\max} + \sigma]$ define domain of discourse, where $D_{\min}$ and $D_{\max}$ are the minimum and maximum inspected value by using standard deviation σ of the data at University of Alabama enrolments (Table 1). In the present study, $\sigma = 1775$ such that $X = [11, 280, 21, 112]$.

Table 1 Time series enrolments data of University of Alabama

Year	Enrolment	Year	Enrolment	Year	Enrolment
1971	13,055	1979	16,807	1987	16,859
1972	13,563	1980	16,919	1988	18,150
1973	13,867	1981	16,388	1989	18,970
1974	14,696	1982	15,433	1990	19,328
1975	15,460	1983	15,497	1991	19,337
1976	15,311	1984	15,145	1992	18,876
1977	15,603	1985	15,163		
1978	15,861	1986	15,984		

Step 2: Domain of discourse X is partitioned into m intervals $e_j(j = 1, 2, \ldots, m)$ of equal length. Fuzzify time series data by employing FSs $F_j(j = 1, 2, \ldots, m)$ with triangular membership function in accordance with the interval $e_j(j = 1, 2, \ldots, m)$.

We construct fourteen fuzzy sets $F_1, F_2, F_3, \ldots, F_{14}$ with equal length interval $e_j(j = 1, 2, 3, \ldots, 14)$ of domain of discourse. Parameters of fourteen triangular fuzzy set $F_j(j = 1, 2, 3, \ldots, 14)$ are defined as follows:

$$F_1 = [11280, 11982.52, 12684.77], \quad F_2 = [11982.52, 12684.77, 13387.01],$$
$$F_3 = [12684.77, 13387.01, 14089.26], \quad F_4 = [13387.01, 14089.26, 14791.51],$$
$$F_5 = [14089.26, 14791.51, 15493.75], \quad F_6 = [14791.51, 15493.75, 16196],$$
$$F_7 = [15493.75, 16196, 16898.25], \quad F_8 = [16196, 16898.25, 17600.49],$$
$$F_9 = [16898.25, 17600.49, 18302.74], \quad F_{10} = [17600.49, 18302.74, 19004.99],$$
$$F_{11} = [18302.74, 19004.99, 19707.23], \quad F_{12} = [19004.99, 19707.23, 20409.47],$$
$$F_{13} = [19707.23, 20409.47, 21112], \quad F_{14} = [20409.47, 21112, 21112].$$

Step 3: Calculate the membership grades of each enrolment from equal intervals using triangular membership function. Compute the probability of membership grades using the following formula.

$$p(\mu_i) = \begin{cases} \frac{l}{\sqrt{2\pi}\zeta_j}\left(e^{-\frac{(x_i-(\mu_i-1)l_j-m_j)^2}{2\zeta_j^2}} + e^{-\frac{(x_i-(1-\mu_i)l_j-m_j)^2}{2\zeta_j^2}}\right); & \mu_i \in [0, 1], \quad 0 \leq p(\mu_i) \leq 1 \\ 0; & \text{otherwise} \end{cases}$$

$$(4)$$

where x_i is the data point with corresponding membership grade μ_i and l_j, m_j, ζ_j are the length, mean, std. deviation of the FSs, respectively, and construct the PFSs. Table 2 shows the probabilistic fuzzy elements (PFEs) of initial five enrolments.

Step 4: An algorithm for the fuzzified data using PFSs along with PFEs as membership grades is as follows:

for i= 1 to n (end of TS data)
    for j= 1 to m (end of intervals)
      choose

$$\mu_{ki} = \max\left(\mu(x_1), \mu(x_2), \ldots, \mu(x_k), \ldots, \mu(x_m)\right), 1 \leq k \leq m$$

        If F'$_k$ is probabilistic fuzzy set corresponds to μ_{ki} then assign F'$_k$ to x$_i$.
      end if
    end for

    end for

Table 2 PFEs of initial five enrolment data

PFSs\Enrol.	13,055	13,563	13,867	14,696	15,460
F_1'	<0|0>	<0|0>	<0|0>	<0|0>	<0|0>
F_2'	<0.4728|0.701>	<0|0>	<0|0>	<0|0>	<0|0>
F_3'	<0.5272|0.739>	<0.7494|0.894>	<0.3165|0.609>	<0|0>	<0|0>
F_4'	<0|0>	<0.2506|0.579>	<0.6835|0.851>	<0.136|0.541>	<0|0>
F_5'	<0|0>	<0|0>	<0|0>	<0.864|0.951>	<0.0481|0.521>
F_6'	<0|0>	<0|0>	<0|0>	<0|0>	<0.9519|0.974>
F_7'	<0|0>	<0|0>	<0|0>	<0|0>	<0|0>
F_8'	<0|0>	<0|0>	<0|0>	<0|0>	<0|0>
F_9'	<0|0>	<0|0>	<0|0>	<0|0>	<0|0>
F_{10}'	<0|0>	<0|0>	<0|0>	<0|0>	<0|0>
F_{11}'	<0|0>	<0|0>	<0|0>	<0|0>	<0|0>
F_{12}'	<0|0>	<0|0>	<0|0>	<0|0>	<0|0>
F_{13}'	<0|0>	<0|0>	<0|0>	<0|0>	<0|0>
F_{14}'	<0|0>	<0|0>	<0|0>	<0|0>	<0|0>
Fuzzified enrolments	F_3'	F_3'	F_4'	F_5'	F_6'

Step 5: FLRs are established employing the rule as follows: For FLR $F_i' \rightarrow F_j'$, here, F_i' is the fuzzy construction of the n year as current state and F_j' is the fuzzy construction of the $n +1$ year as next state. Fuzzified time series data of enrolments and applying an algorithm for fuzzification (step 4), FLRs and FLRGs (Table 3) are determined.

Use max–min composition operations on FLR to have upshots associated with the probabilities to construct the fuzzy row vector by the following aggregating operator.

$$f_i = \left(1 - (1 - o_i)^{p_i}\right)^{1/p_i}, \quad 0 \le p_i \le 1 \tag{5}$$

where p_i is the probability of corresponding output o_i and f_i is the probabilistic fuzzified upshots and defuzzify by centroid formula for the numerical forecast as follows.

$$\text{Numerical forecast} = \frac{\sum_{i=1}^{n} f_i c_i}{\sum_{i=1}^{n} f_i} \tag{6}$$

where c_i is the centroid for the equal intervals.

Table 3 FLRs and FLRGs for the historical data

FLRs		FLRGs
$F'_3 \rightarrow F'_3$	$F'_3 \rightarrow F'_4$	$F'_3 \rightarrow F'_3, F'_4$
$F'_6 \rightarrow F'_6$	$F'_7 \rightarrow F'_8$	$F'_4 \rightarrow F'_5$
$F'_7 \rightarrow F'_6$	$F'_4 \rightarrow F'_5$	$F'_5 \rightarrow F'_6$
$F'_6 \rightarrow F'_8$	$F'_{10} \rightarrow F'_{11}$	$F'_6 \rightarrow F'_6, F'_7$
$F'_{11} \rightarrow F'_{11}$	$F'_5 \rightarrow F'_6$	$F'_7 \rightarrow F'_6, F'_8$
$F'_6 \rightarrow F'_6$	$F'_8 \rightarrow F'_8$	$F'_8 \rightarrow F'_7, F'_8, F'_{10}$
$F'_8 \rightarrow F'_{100}$	$F'_{11} \rightarrow F'_{11}$	$F'_{10} \rightarrow F'_{11}$
$F'_6 \rightarrow F'_6$	$F'_6 \rightarrow F'_6$	$F'_{11} \rightarrow F'_{11}$
$F'_6 \rightarrow F'_7$	$F'_6 \rightarrow F'_7$	
$F'_6 \rightarrow F'_6$	$F'_{11} \rightarrow F'_{11}$	
$F'_8 \rightarrow F'_7$		

4   Performance Analysis

Performance of proposed forecasting method is analysed using different error measures RMSE and AFE which are defined as follows:

$$\text{RMSE} = \sqrt{\frac{\sum_{i=1}^{n} (F_i - O_i)^2}{n}} \tag{7}$$

$$\text{AFE(in\%)} = \frac{1}{n} \sum_{i=1}^{n} \frac{|F_i - O_i|}{O_i} \times 100 \tag{8}$$

where O_i and F_i denote the observed and forecasted TS data, and n is an index of data points. Proposed forecasting method is compared with existing methods suggested by the various researchers [2, 3, 9, 10, 13, 18–20, 25, 27, 31] in terms of RMSE and AFE.

5   Conclusion

Both probabilistic and non-probabilistic uncertainties simultaneously occur in the system. In this study, we have presented a probabilistic FTS forecasting method that includes probabilistic and non-probabilistic uncertainties in time series forecasting. The presented method is compared with few recent FTS forecasting methods. Decreased value of both RMSE and AFE validate its outperformance in forecasting University of Alabama enrolments. Even though both RMSE and AFE are slightly higher than that of models proposed by Joshi and Kumar [18] and Bisht and Kumar [2], its competency in handling both uncertainties in single framework makes it more efficient (Table 4).

Table 4 Forecasted value of university enrolments by proposed and other existing models RMSE and AFER.

Actual Enrolment	Song and Chissom [25]	Chen [3]	Lee and Chou [20]	SC time variant [27]	Cheng et al. [9]	Cheng et al. [10]	Yolcu et al. [31]	Egrioglu [13]	Joshi and Kumar [18]	Kumar and Gangwar [19]	Bisht and Kumar [2]	Proposed method
13,055	–	–	–	–	–	–	–	–	–	–	–	–
13,563	14,000	14,000	14,025	–	14,230	14,242	14031.35	13480.63	14,250	–	13595.67	13559.71
13,867	14,000	14,000	14,568	–	14,230	14,242	14795.36	13480.63	14,246	13,963	13814.75	13559.71
14,696	14,000	14,000	14,568	14,700	14,230	14,242	14795.36	14,242	14,246	13,963	14929.79	14781.24
15,460	15,500	15,500	15,654	14,800	15,541	15474.3	14795.36	15,710	15,491	14,867	15541.27	15333.51
15,311	16,000	16,000	15,654	15,400	15,541	15474.3	16406.57	15484.55	15,491	15,287	15540.62	15677.52
15,603	16,000	16,000	15,654	15,500	15,541	15474.3	16406.57	15935.65	15,491	15,376	15540.62	15677.52
15,861	16,000	16,000	15,654	15,500	16,196	15474.3	16406.57	15935.65	16,345	15,376	15540.62	15677.52
16,807	16,000	16,000	16,197	16,800	16,196	16146.5	16406.57	16837.86	16,345	15,376	16254.5	15965.56
16,919	16,813	16,833	17,283	16,200	16,196	16988.3	17315.29	17499.69	15,850	16,523	17040.41	17001.21
16,388	16,813	16,833	17,283	16,400	17,507	16988.3	17315.29	17499.69	15,850	16,606	17040.41	17001.21
15,433	16,789	16,833	16,197	16,800	16,196	16146.5	17315.29	16,737	15,850	17,519	16254.5	15965.56
15,497	16,000	16,000	15,654	16,400	15,541	15474.3	16406.57	15484.55	15,450	16,606	15540.62	15677.52
15,145	16,000	16,000	15,654	15,500	15,541	15474.3	16406.57	15484.55	15,450	15,376	15540.62	15677.52
15,163	16,000	16,000	15,654	15,500	15,541	15474.3	16406.57	15,710	15,491	15,376	15541.27	15677.52
15,984	16,000	16,000	15,654	15,500	15,541	15474.3	16406.57	15,710	15,491	15,287	15541.27	15677.52
16,859	16,000	16,000	16,197	16,800	16,196	16146.5	16406.57	16837.86	16,345	15,287	16254.5	15965.56
18,150	16,813	16,833	17,283	19,300	17,507	16988.3	17315.29	17499.69	17,950	16,523	17040.41	17001.21
18,970	19,000	19,000	18,369	17,800	18,872	19,144	19132.79	19144.4	18,961	17,519	18902.3	19159.7
19,328	19,000	19,000	19,454	19,300	18,872	19,144	19132.79	19144.4	18,961	19,500	19357.3	19132.11
19,337	19,000	19,000	19,454	19,600	18,872	19,144	19132.79	19144.4	18,961	19,000	19168.56	19132.11
18,876	–	19,000	–	–	18,872	19,144	19132.79	19144.4	18,961	19,500	19168.56	19132.11
RMSE	650.4	638.36	501.28	880.73	511.04	478.45	805.17	484.61	433.76	493.56	428.63	470.08
AFE (%)	3.22	3.11	2.67	3.75	2.66	2.39	4.29	2.21	2.24	2.33	1.94	2.21

References

1. Almeida, R.J., Kaymak, U.: Probabilistic fuzzy systems in value-at-risk estimation. Intell. Syst. Acc. Finan. Manag. **16**(1–2), 49–70 (2009)
2. Bisht, K., Kumar, S.: Fuzzy time series forecasting method based on hesitant fuzzy sets. Expert Syst. App. **64**, 557–568 (2016)
3. Chen, S.M.: Forecasting enrolments based on fuzzy time series. Fuzzy Sets Syst. **81**(3), 311–319 (1996)
4. Chen, S.M., Hwang, J.R.: Temperature prediction using fuzzy time series. IEEE Trans. Syst. Man Cyber. Part B (Cybernetics) **30**(2), 263–275 (2000)
5. Chen, S.M., Tanuwijaya, K.: Multivariate fuzzy forecasting based on fuzzy time series and automatic clustering techniques. Expert Syst. App. **38**(8), 10594–10605 (2011)
6. Chen, S.M., Kao, P.Y.: TAIEX forecasting based on fuzzy time series, particle swarm optimization techniques and support vector machines. Inf. Sci. **247**, 62–71 (2013)
7. Chen, M.Y., Chen, B.T.: A hybrid fuzzy time series model based on granular computing for stock price forecasting. Inf. Sci. **294**, 227–241 (2015)
8. Chen, S.M., Phuong, B.D.H.: Fuzzy time series forecasting based on optimal partitions of intervals and optimal weighting vectors. Knowledge-Based Syst. **118**, 204–216 (2017)
9. Cheng, C.H., Chang, J.R., Yeh, C.A.: Entropy-based and trapezoid fuzzification-based fuzzy time series approaches for forecasting IT project cost. Tech. Forec. Soc. Change **73**(5), 524–542 (2006)
10. Cheng, C.H., Cheng, G.W., Wang, J.W.: Multi-attribute fuzzy time series method based on fuzzy clustering. Expert Syst. App. **34**(2), 1235–1242 (2008)
11. Cheng, S.H., Chen, S.M., Jian, W.S.: Fuzzy time series forecasting based on fuzzy logical relationships and similarity measures. Inf. Sci. **327**, 272–287 (2016)
12. Deng, W., Wang, G., Zhang, X., Xu, J., Li, G.: A multi-granularity combined prediction model based on fuzzy trend forecasting and particle swarm techniques. Neuro comp. **173**, 1671–1682 (2016)
13. Eğrioğlu, E.: A new time-invariant fuzzy time series forecasting method based on genetic algorithm. Adv. Fuzzy Syst. **2** (2012)
14. Fialho, A.S., Vieira, S.M., Kaymak, U., Almeida, R.J., Cismondi, F., Reti, S.R., Sousa, J.M.: Mortality prediction of septic shock patients using probabilistic fuzzy systems. App. Soft Comp. **42**, 194–203 (2016)
15. Hinojosa, W.M., Nefti, S., Kaymak, U.: Systems control with generalized probabilistic fuzzy-reinforcement learning. IEEE Trans. Fuzzy Syst. **19**(1), 51–64 (2011)
16. Huarng, K.: Effective lengths of intervals to improve forecasting in fuzzy time series. Fuzzy Sets Syst. **123**(3), 387–394 (2001)
17. Huarng, K., Yu, T.H.K.: Ratio-based lengths of intervals to improve fuzzy time series forecasting. IEEE Trans. Syst. Man Cyber. Part B (Cybernetics) **36**(2), 328–340 (2006)
18. Joshi, B.P., Kumar, S.: Intuitionistic fuzzy sets based method for fuzzy time series forecasting. Cyber. Syst. **43**(1), 34–47 (2012)
19. Kumar, S., Gangwar, S.S.: Intuitionistic fuzzy time series: an approach for handling non-determinism in time series forecasting. IEEE Trans. Fuzzy Syst. **24**(6), 1270–1281 (2016)
20. Lee, H.S., Chou, M.T.: Fuzzy forecasting based on fuzzy time series. Int. J. Comput. Math. **81**(7), 781–789 (2004)
21. Li Y, Huang W (2012) A Probabilistic fuzzy set for uncertainties-based modeling in logistics manipulator system. J. Th. App. Inf. Tech. **46**(2)
22. Liu, Z., Li, H.X.: A probabilistic fuzzy logic system for modeling and control. IEEE Trans. Fuzzy Syst. **13**(6), 848–859 (2005)
23. Liu, H.T.: An improved fuzzy time series forecasting method using trapezoidal fuzzy numbers. Fuzzy Opti. Deci. Mak. **6**(1), 63–80 (2007)

24. Meghdadi, A.H., Akbarzadeh-T,M.R.: Probabilistic fuzzy logic and probabilistic fuzzy systems. In The 10th IEEE International Conference on Fuzzy Systems, 2001 vol. 3, pp. 1127–113. IEEE (2001)
25. Song, Q., Chissom, B.S.: Fuzzy time series and its models. Fuzzy Sets Syst. **54**(3), 269–277 (1993)
26. Song, Q., Chissom, B.S.: Forecasting enrolments with fuzzy time series—part I. Fuzzy Sets Syst. **54**(1), 1–9 (1993)
27. Song, Q., Chissom, B.S.: Forecasting enrolments with fuzzy time series—part II. Fuzzy Sets Syst. **62**(1), 1–8 (1994)
28. Song, Q.: A note on fuzzy time series model selection with sample autocorrelation functions. Cyber. Syst. **34**(2), 93–107 (2003)
29. Xian, S., Zhang, J., Xiao, Y., Pang, J.: A novel fuzzy time series forecasting method based on the improved artificial fish swarm optimization algorithm. Soft Comp. 1–11 (2017)
30. Ye, F., Zhang, L., Zhang, D., Fujita, H., Gong, Z.: A novel forecasting method based on multi-order fuzzy time series and technical analysis. Inf. Sci. **367**, 41–57 (2016)
31. Yolcu, U., Egrioglu, E., Uslu, V.R., Basaran, M.A., Aladag, C.H.: A new approach for determining the length of intervals for fuzzy time series. App. Soft Comp. **9**(2), 647–651 (2009)
32. Yolcu, O.C., Yolcu, U., Egrioglu, E., Aladag, C.H.: High order fuzzy time series forecasting method based on an intersection operation. App. Math. Mod. **40**(19), 8750–8765 (2016)
33. Zadeh, L.A.: Fuzzy sets. Inf. Cont. **8**(3), 338–353 (1965)

Ohmic–Viscous Dissipation and Heat Generation/Absorption Effects on MHD Nanofluid Flow Over a Stretching Cylinder with Suction/Injection

Ashish Mishra and Manoj Kumar

Abstract This article classifies the influence of ohmic–viscous dissipation and heat generation/absorption on MHD flow of Ag–water nanofluid over a stretching cylinder in the occurrence of suction/injection. The set of obtained ODEs have been explained along with assisting boundary conditions by employing nonlinear numerical approach called Runge–Kutta–Fehlberg scheme through shooting procedure. The impact of pertinent factors on non-dimensional flow and thermal profiles is shown by graphs and explained in detail. Also, the dimensionless heat transfer rate is established in tabular way. The outcomes validated that as values of thermal slip and Eckert number accelerated, the heat transfer coefficient lessens while it enhances with a rise in the suction/injection parameter.

Keywords Heat generation/absorption · Nanofluid · Ohmic dissipation
Stretching cylinder · Suction/injection · Viscous dissipation

1 Introduction

During the last few decades, the scrutiny of boundary layer flow of viscous fluids stimulated by stretching sheet has massive functions in manufacturing procedures and polymer production. Applications of such industrial techniques concerning polymers contain cooling of filaments, paper productions, glass blowing, synthetic filaments, bar drawing, hot rolling, fibreglass, aerodynamics, etc. Wang [1] premeditated the steady fluid flow past a stretching tube from external plane. Ishak et al. [2] simulated impact of suction or injection on steady stream of an incompressible liquid towards permeable stretching cylinder. Ahmed et al. [3] described

A. Mishra (✉) · M. Kumar
Department of Mathematics, Statistics and Computer Science, G. B. Pant University
of Agriculture and Technology, Pantnagar 263145, India
e-mail: ashushmishra@gmail.com

M. Kumar
e-mail: mnj_kumar2004@yahoo.com

© Springer Nature Singapore Pte Ltd. 2019
J. K. Mandal et al. (eds.), *Advanced Computing and Communication Technologies*,
Advances in Intelligent Systems and Computing 702,
https://doi.org/10.1007/978-981-13-0680-8_5

the joined consequence of thermal conductivity and dynamic viscosity in existence of stretching permeable tube in nanofluid. Ashorynejad et al. [4] have scrutinized the impact of magnetohydrodynamic stream over a stretching tube within nanofluid. Majeed et al. [5] have conferred about the combined effect of heat transfer with partial slip on steady non-Newtonian flow of Casson fluid on stretching pipe managed with heat flux. Hayat et al. [6] investigated MHD third-grade fluid stream caused by a stretching cylindrical plane. The impact of Joule heating and viscous dissipation past an axially stretching cylinder loaded with MHD Sisko nanofluid was explored by Hussain et al. [7]. MHD boundary layer flow of slip along with an elongate cylinder was scrutinized by Mukhopadhyay [8]. Pandey and Kumar [9] employed RKF procedure to explain the principal equations of heat transfer of copper–water nanofluid towards a stretched tube in occurrence of slip and viscous dissipation. Malik et al. [10] proposed influence of free convection past an upright stretching cylinder in existence of Casson nanofluid. Abbas et al. [11] evaluated the influences of slip on the stream over an unsteady stretching/shrinking cylinder within suction. The KBM was utilized by Majeed et al. [12] to examine the Soret and Dufour impacts with thermal radiation on second-grade fluid caused by stretching cylinder. Ishak and Nazar [13] considered incompressible and laminar flow of viscous fluid caused by cylinder. Vajravelu et al. [14] explored heat transfer tendency past a permeable stretching tube by considering temperature-based thermo-physical attributes. Malik et al. [15] proposed MHD hyperbolic fluid stream over stretched plane. The varieties of investigations to inspect the impact of ohmic–viscous dissipation on flow of nanofluid under assorted conditions have been proposed by [16, 17].

The main objective of the paper is to achieve the numerical elucidation of silver–water nanofluid flow over a stretching cylinder by taking into consideration the influences of ohmic–viscous dissipation, velocity and temperature slips, suction/blowing and heat generation/absorption.

2 Mathematical Problem

Consider an axisymmetric, incompressible, steady, laminar flow of a nanofluid over a flat stretching cylinder of radius a and (z, r) are taken as directions of axis towards horizontal and vertical of the cylinder as depicted in Fig. 1. We also assumed that B_0 is intensity of uniform magnetic field towards radial direction. It is supposed that surface temperature of pipe is T_w and temperature far from the surface is T_∞, where $(T_w - T_\infty) > 0$. Also, physical features of solid particles and regular fluid are mentioned in Table 1. The primary equations of mass, momentum and energy are articulated as (Refs. [2–4]):

Fig. 1 Physical model and coordinate system

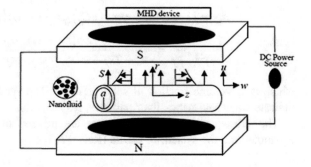

Table 1 Thermophysical properties of pure water and Ag nanoparticle (Ref. [3])

	$\rho\left(\text{kg/m}^3\right)$	$C_p(\text{J/kg K})$	$k(\text{W/m K})$	$\sigma(\text{S/m})$
Pure water	997.1	4179	0.613	0.05
Silver (Ag)	10,500	235	429	6.3×10^7

$$\frac{\partial(rw)}{\partial z} + \frac{\partial(ru)}{\partial r} = 0 \tag{1}$$

$$\rho_{nf}\left(w\frac{\partial w}{\partial z} + u\frac{\partial w}{\partial r}\right) = \mu_{nf}\left(\frac{\partial^2 w}{\partial z^2} + \frac{1}{r}\frac{\partial w}{\partial r}\right) - \sigma_{nf}B_0^2 w \tag{2}$$

$$\rho_{nf}\left(w\frac{\partial u}{\partial z} + u\frac{\partial u}{\partial r}\right) = -\frac{\partial p}{\partial r} + \mu_{nf}\left(\frac{\partial^2 u}{\partial r^2} + \frac{1}{r}\frac{\partial u}{\partial r} - \frac{u}{r^2}\right) \tag{3}$$

$$\left(\rho C_p\right)_{nf}\left(w\frac{\partial T}{\partial z} + u\frac{\partial T}{\partial r}\right) = \kappa_{nf}\left(\frac{\partial^2 T}{\partial r^2} + \frac{1}{r}\frac{\partial T}{\partial r}\right) + Q_0(T - T_\infty) + \mu_{nf}\left(\frac{\partial w}{\partial r}\right)^2 + \sigma_{nf}B_0^2 w^2 \tag{4}$$

The associated boundary conditions are as follows (Refs. [2–4]):

$$\left.\begin{aligned} u = U_w, w = W_w + l_1\frac{\partial w}{\partial r}, T = T_w + l_2\frac{\partial T}{\partial r} \text{ at } r = a \\ u \to 0, T \to T_\infty \text{ as } r \to \infty \end{aligned}\right\} \tag{5}$$

where l_1 is velocity slip factor and l_2 is thermal slip factor. The effective density ρ_{nf}, heat capacitance $\left(\rho C_p\right)_{nf}$, dynamic viscosity μ_{nf}, thermal conductivity κ_{nf} and electric conductivity σ_{nf} of nanofluid are described as (Refs. [4, 17]):

$$\left.\begin{array}{l} \rho_{nf} = (1-\varphi)\rho_f + \varphi\rho_s,\, (\rho C_p)_{nf} = (1-\varphi)(\rho C_p)_f + \varphi(\rho C_p)_s, \\ \dfrac{\mu_{nf}}{\mu_f} = \dfrac{1}{(1-\varphi)^{2.5}},\, \dfrac{\kappa_{nf}}{\kappa_f} = \dfrac{\kappa_s + 2\kappa_f - 2\varphi(\kappa_f - \kappa_s)}{\kappa_s + 2\kappa_f + \varphi(\kappa_f - \kappa_s)},\, \dfrac{\sigma_{nf}}{\sigma_f} = (1-\varphi) + \varphi\dfrac{\sigma_s}{\sigma_f} \end{array}\right\} \quad (6)$$

where φ is volume fraction of solid particles. The symbols s and f in subscripts indicate water and nano-silver particles, respectively.

The subsequent similarity variables are utilized to transform the governing equations into the system of ODEs (Refs. [2–4]):

$$u = -\frac{ca}{\sqrt{\eta}}f(\eta), w = 2zcf'(\eta), \eta = \left(\frac{r}{a}\right)^2, \theta(\eta) = \frac{T - T_\infty}{T_w - T_\infty}. \quad (7)$$

Substituting Eqs. (6) and (7) into Eqs. (2) and (4), we get consequent system of nonlinear ODEs:

$$\frac{\mu_{nf}}{\mu_f}[\eta f''' + f''] - MA_5f' + A_1 Re[ff'' - f'^2] = 0 \quad (8)$$

$$RePrA_2f\theta' + \frac{\mu_{nf}}{\mu_f}(\eta PrEc)f''^2 + \frac{\kappa_{nf}}{\kappa_f}(\eta\theta'' + \theta') + Q\theta + MEcPrA_5f'^2 = 0 \quad (9)$$

The transformed boundary conditions in the terms of f and θ are explained as follows:

$$\left.\begin{array}{l} f(1) = S, f'(1) = 1 + \text{Vs}\, f''(1), \theta(1) = 1 + \text{Ts}\,\theta'(1) \text{ at } \eta = 1 \\ f'(\infty) \to 0, \theta(\infty) \to 0 \text{ as } \eta \to \infty. \end{array}\right\} \quad (10)$$

where prime indicates derivative with respect to η. The dimensionless existing parameters are namely S suction $(S > 0)$ and injection $(S < 0)$, Vs velocity slip, Ts thermal slip, Pr Prandtl number, M magnetic field, Re Reynolds number, Q heat generation/absorption, Ec Eckert number and A_1, A_2, A_3, A_4 and A_5 are constants, respectively. These significant parameters dictating the flow dynamics are defined as follows:

$$\left.\begin{array}{l} \text{Vs} = \dfrac{2l_1}{a}, \text{Ts} = \dfrac{2l_2}{a}, Pr = \dfrac{\nu_f}{\alpha_f}, M = \dfrac{\sigma_f B_0^2 a^2}{4\rho_f \nu_f}, Re = \dfrac{ca^2}{2\alpha_f}, S = -\dfrac{\sqrt{\eta}U_w}{ca}, \\ Q = \dfrac{Q_0 a^2}{4\kappa_f}, Ec = \dfrac{4\rho_f c^2 z^2}{(\rho C_p)_f(T_w - T_\infty)}, A_1 = (1-\varphi) + \varphi\left(\dfrac{\rho_s}{\rho_f}\right), \\ A_2 = (1-\varphi) + \varphi\left(\dfrac{(\rho C_p)_s}{(\rho C_p)_f}\right), A_3 = \dfrac{\mu_{nf}}{\mu_f}, A_4 = \dfrac{\kappa_{nf}}{\kappa_f}, A_5 = \dfrac{\sigma_{nf}}{\sigma_f} \end{array}\right\} \quad (11)$$

For the practical concern the significant quantities, the skin friction coefficient and the local Nusselt number are delineated as follows:

$$C_f = \frac{2\mu_{nf}}{W_w^2 \rho_f} \left(\frac{\partial w}{\partial r} \right)_{r=a}, Nu = -\frac{a\kappa_{nf}}{\kappa_f (T_w - T_\infty)} \left(\frac{\partial T}{\partial r} \right)_{r=a} \tag{12}$$

Now, using Eqs. (6) and (7) in Eq. (12), the dimensionless skin friction coefficient and Nusselt number are developed into non-dimensional reduced skin friction coefficient and reduced Nusselt number, which can be articulated as follows:

$$(zRe/a)C_f = \frac{\mu_{nf}}{\mu_f} f''(1), Nu = -\frac{2\kappa_{nf}}{\kappa_f (T_w - T_\infty)} \theta'(1) \tag{13}$$

3 Numerical Method

The dimensionless momentum Eq. (8) and energy Eq. (9) together with associated boundary conditions (10) have been elucidated numerically by applying Runge–Kutta–Fehlberg technique of 4–5th order with shooting procedure. In this problem, we initially set up the following terms into the obtained boundary layer equations to turn it into first-order ODEs.

Let us consider $y_1 = f$, $y_2 = f'$, $y_3 = f''$, $y_4 = \theta$, $y_5 = \theta'$.

Using above mentioned substitutions in such a manner that acquired ODEs are transformed into an arrangement of nonlinear first-order ODEs. Now the following first-order system of ODEs is obtained as follows:

$$\begin{pmatrix} y_1' \\ y_2' \\ y_3' \\ y_4' \\ y_5' \end{pmatrix} = \begin{pmatrix} y_2 \\ y_3 \\ \frac{A_1 Re(y_2 y_2 - y_1 y_3) - A_3 y_3 + A_5 M y_2}{\eta A_3} \\ y_5 \\ \frac{-A_4 y_5 - A_2 Pr Re\, y_1 y_5 - Q y_4 - A_3 \eta Pr Ec\, y_3 y_3 - A_5 EcM Pr y_2 y_2}{\eta A_4} \end{pmatrix} \tag{14}$$

and subsequent initial conditions:

$$(y_1, y_2, y_3, y_4, y_5)^T = (S, 1 + pVs, p, 1 + qTs, q)^T \tag{15}$$

The system of first-order ODEs (14) together with preliminary conditions (15) is explained with the aid of order of fourth and fifth RKF integration procedure. Here, we contrast the evaluated values of $f'(\eta)$ and $\theta(\eta)$ as $\eta = 1$, during specified boundary conditions $f'(\eta \to \infty) = 0$ and $\theta(\eta \to \infty) = 0$. The unidentified constants p and q have been approximated by Newton's scheme in such a manner that boundary conditions suited at large numerical values of $\eta \to \infty$ with error below 10^{-6}.

4 Results and Discussion

The numeric explanations are achieved for velocity and temperature profiles as a result of several significances of acting parameters which are displayed in Figs. 2, 3, 4, 5, 6, 7, 8 and 9. We have contrasted the results of $(-\theta'(1))$ for Pr with those found by Wang [1], Ishak et al. [2] and Ahmed et al. [3] in case of flat surface. A close agreement is found on viewing Table 2. The skin friction coefficient and heat transfer rate are revealed in Table 3 for diverse values of Vs, Ts, M, Ec, Q and S of Ag nanoparticles. Throughout the whole process for flow analysis, the value of parameter like $Pr = 6.2$ and $Re = 5$ is fixed, when $1 \leq \eta \leq 6$ with preset mesh size

Fig. 2 Variation in velocity profiles due to Vs

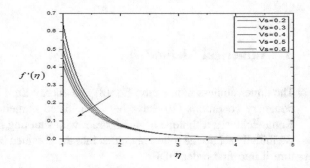

Fig. 3 Variation in temperature profiles due to Ts

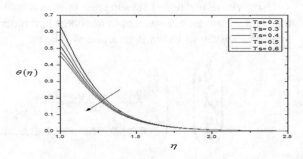

Fig. 4 Variation in velocity profiles due to M

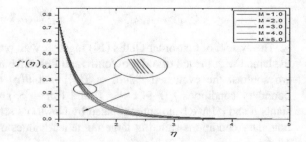

Fig. 5 Variation in
temperature profiles due to M

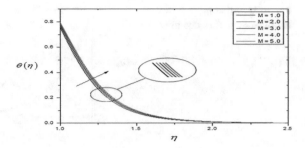

Fig. 6 Variation in
temperature profiles due to Ec

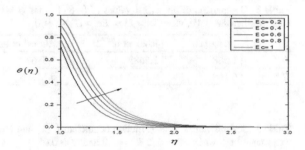

Fig. 7 Variation in velocity
profiles due to S

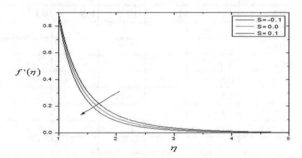

Fig. 8 Variation in
temperature profiles due to S

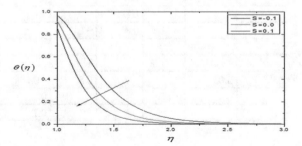

Fig. 9 Variation in
temperature profiles due to Q

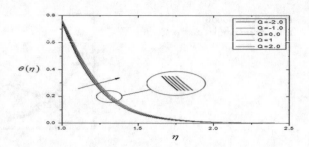

Table 2 Comparison of numerical values of $(-\theta'(1))$, for $\varphi = 0$ with various values of Prandtl number when $Re = 10, Vs = Ts = M = Ec = Q = S = 0$.

Pr	Wang [1]	Ishak et al. [2]	Ahmed et al. [3]	Present study
0.7	1.568	1.5683	1.58679	1.586790
7	6.160	6.1592	6.15776	6.155812

Table 3 Several values of skin friction coefficient $f''(1)$ and Nusselt number $(-\theta'(1))$ for Ag–water nanofluid, when $Pr = 6.2$, $Re = 5.0$ and $\varphi = 0.05$

Vs	Ts	M	Ec	Q	S	$-f''(1)$	$-\theta'(1)$
0.2	0.1	1	0.3	0.5	0.1	1.758939	2.596423
0.3						1.456033	2.669407
0.4						1.246727	2.700071
0.1	0.2					2.243817	1.816193
	0.3					2.243817	1.452303
	0.4					2.243817	1.209914
	0.1	2				2.310603	2.320517
		3				2.373251	2.225806
		4				2.432367	2.137641
		1	0.4			2.243815	2.116847
			0.6			2.243815	1.503545
			0.8			2.243815	0.890599
			0.3	−1		2.243815	2.536324
				0		2.243815	2.462111
				1		2.243815	2.383643
0.05	0.05			0.5	−0.1	2.151747	0.607951
					0	2.375904	1.593203
					0.1	2.621011	2.713308

$\Delta \eta = 0.001$. The variation in profiles of dimensionless field $f'(\eta)$ for some elected values of velocity slip parameter Vs is disclosed in Fig. 2, when Ts $= 0.1$, $Ec = 0.3$, $Q = 0.5$, $M = 10^{-8}$, $S = 0.1$ and $\varphi = 0.05$. From this outline, it is alleged that on enhancing the values of Vs, velocity of nanofluid continuously diminishes. Figure 3 shows the alteration in temperature graph corresponding to variable η, for numerous values of thermal slip parameter Ts. From this outline, it is perceived that temperature of nanofluid is a decreasing function of Ts. Moreover, Table 3 shows that heat transfer rate reduces with a rise in Ts. Figure 4 exhibits that velocity profile of nanofluid diminishes with an increase in magnetic parameter. It is because of the piece of information that the occurrence regarding the magnetic field in movement generates a force determined as the Lorentz force which performs as an impeded force. Similarly, as referring to Fig. 5, we analysed that thermal field varies with magnetic field parameter M corresponding to each value of horizontal component η in the realm of $[1, 2.5]$ and the trend of temperature profiles augment with a boost in M for entire specified region. Figure 6 illustrates the sketch between temperature profiles versus similarity variable η for diverse values of Eckert number Ec. It is obvious from this graph that temperature of nanoparticles enhances with a rise in Eckert number. Moreover, Table 3 reveals that value of Nusselt number is exponentially decreased as augmentation in viscous dissipation parameter Ec is done. Figure 7 presents the velocity distribution against the variable η for preferred values of suction/injection parameter S; from this outline, we deduce that when parameter S shifts their position from injection to suction then velocity profiles decrease in the region $1 \leq \eta < 5$. Consequently, the outline of velocity boundary layer thickness continuously becomes thinner with magnification in suction/injection parameter. Figure 8 displays the strong impact of suction/injection on the temperature fields. These curves identified that thermal field continuously decreases, as parameter S moves from injection to suction domain. The effect of suction/injection on skin friction coefficient is revealed in Table 3. On viewing this table, we elucidate that the value of $-f''(1)$ exponentially increases along with S. Figure 9 has been drawn to scrutinize the temperature profiles of nanofluid corresponding to variable η, for assorted values of Q. It is discerned that temperature profile approaches to zero asymptotically in the range of $1 \leq \eta < \infty$. In other words, it enhances continuously with elevation in heat generation/absorption Q in the dynamic range of $[1, \infty)$. Furthermore, the thermal boundary layer width continuously enhances. However, heat transfer rate regularly diminishes, with an increment in Q.

5 Conclusions

In this article, the influence of viscous–ohmic dissipation, suction or injection with heat generation/absorption on two-dimensional steady MHD boundary layer flow of water-based nanofluid containing Ag (silver) nanoparticles over a stretching cylinder in existence of slip boundary conditions have been discussed. The major conclusions of this work are as follows:

- The temperature field of Ag–water nanofluid boosts with enhancement in the magnetic field, heat generation/absorption in addition to Eckert number.
- Escalating values of velocity slip reduce the velocity, while thermal slip decreases the temperature.
- On escalating the magnetic field, the momentum boundary layer width reduces.
- Nusselt number is diminished with escalating values of magnetic field, heat generation/absorption parameter and Eckert number. However, it illustrates reverse trend when velocity slip and suction/injection are escalated.

The obtained results have huge significance in different fields of science and technology, where layers of the surface are being stretched.

References

1. Wang, C.Y.: Fluid flow due to a stretching cylinder. Phys. Fluid. **31**(3), 466–468 (1988)
2. Ishak, A., Nazar, R., Pop, I.: Uniform suction/blowing effect on flow and heat transfer due to a stretching cylinder. Appl. Math. Model. **32**(10), 2059–2066 (2008)
3. Ahmed, S.E., Hussein, A.K., Mohammed, H.A., Sivasankaran, S.: Boundary layer flow and heat transfer due to permeable stretching tube in the presence of heat source/sink utilizing nanofluids. Appl. Math. Comput. **238**, 149–162 (2014)
4. Ashorynejad, H.R., Sheikholeslami, M., Pop, I., Ganji, D.D.: Nanofluid flow and heat transfer due to a stretching cylinder in the presence of magnetic field. Heat Mass Transf. **49**(3), 427–436 (2013)
5. Majeed, A., Javed, T., Ghaffari, A., Rashidi, M.M.: Analysis of heat transfer due to stretching cylinder with partial slip and prescribed heat flux: a Chebyshev spectral Newton iterative scheme. Alexandria Eng. J. **54**(4), 1029–1036 (2015)
6. Hayat, T., Shafiq, A., Alsaedi, A.: MHD axisymmetric flow of third grade fluid by a stretching cylinder. Alexandria Eng. J. **54**(2), 205–212 (2015)
7. Hussain, A., Malik, M.Y., Salahuddin, T., Bilal, S., Awais, M.: Combined effects of viscous dissipation and Joule heating on MHD Sisko nanofluid over a stretching cylinder. J. Mol. Liq. **231**, 341–352 (2017)
8. Mukhopadhyay, S.: MHD boundary layer slip flow along a stretching cylinder. Ain Shams Eng. J. **4**(2), 317–324 (2013)
9. Pandey, A.K., Kumar, M.: Natural convection and thermal radiation influence on nanofluid flow over a stretching cylinder in a porous medium with viscous dissipation. Alexandria Eng. J. **56**(1), 55–62 (2017)
10. Malik, M.Y., Naseer, M., Nadeem, S., Rehman, A.: The boundary layer flow of Casson nanofluid over a vertical exponentially stretching cylinder. Appl. Nanosci. **4**(7), 869–873 (2014)
11. Abbas, Z., Rasool, S., Rashidi, M.M.: Heat transfer analysis due to an unsteady stretching/shrinking cylinder with partial slip condition and suction. Ain Shams Eng. J. **6**(3), 939–945 (2015)
12. Majeed, A., Javed, T., Ghaffari, A.: Numerical investigation on flow of second grade fluid due to stretching cylinder with Soret and Dufour effects. J. Mol. Liq. **221**, 878–884 (2016)
13. Ishak, A., Nazar, R.: Laminar boundary layer flow along a stretching cylinder. Eur. J. Sci. Res. **36**(1), 22–29 (2009)

14. Vajravelu, K., Prasad, K.V., Santhi, S.R., Umesh, V.: Fluid flow and heat transfer over a permeable stretching cylinder. J. Appl. Fluid Mech. **7**(1), (2014)
15. Malik, M.Y., Salahuddin, T., Hussain, A., Bilal, S.: MHD flow of tangent hyperbolic fluid over a stretching cylinder: using Keller box method. J. Magn. Magn. Mat. **395**, 271–276 (2015)
16. Pal, D., Mandal, G.: Double diffusive magnetohydrodynamic heat and mass transfer of nanofluids over a nonlinear stretching/shrinking sheet with viscous-Ohmic dissipation and thermal radiation. Propul. Power Res. **6**(1), 58–69 (2017)
17. Pal, D., Mandal, G.: Hydromagnetic convective–radiative boundary layer flow of nanofluids induced by a non-linear vertical stretching/shrinking sheet with viscous–Ohmic dissipation. Powder Technol. **279**, 61–74 (2015)

Interlocking Nodes for Structural Analysis in Social Networking

S. A. S. Bommakanti

Abstract In this work, an algorithm for detecting the interlocking nodes in the temporal networks has been proposed. Interlocking nodes are set of nodes joining together in same set of networks. These nodes make change in the structural changes of the temporal network. Different techniques exist in the literature to identify structural changes of the temporal network. Structural changes are essential elements for identifying patterns and events in temporal social networks. A method for finding structural changes and the events related to communities is presented in the paper. These events can be used for pattern detection in networks with evolving communities.

Keywords Interlocking nodes · Temporal social networks · Evolution

1 Introduction

Social network analysis is the process of investigating social structures through the use of network and graph theories [1]. In SNA, network can be represented as nodes and edges. Nodes can be things, people, or actors. And edges are the relation between the nodes. A few examples are friendship networks, co-authorship between researchers, e-mail interaction between employees within the organization, and CEO connections in corporate networks. An important aspect in most social networks is their ability to evolve over time. This evolution of communities is affected by the influential nodes. The "influential nodes" can be regarded as the most dominating nodes in the social networks, which induce social influence phenomenon (SIP) among the neighboring nodes in that network. Social influence is the phenomenon which causes the actions on a user to influence their friends to behave in a way to follow their behavior in that network [2]. For example, if a

S. A. S. Bommakanti (✉)
Birla Institute of Technology and Science Pilani, Hyderabad Campus,
Hyderabad 500078, India
e-mail: p20150409@hyderabad.bits-pilani.ac.in

© Springer Nature Singapore Pte Ltd. 2019
J. K. Mandal et al. (eds.), *Advanced Computing and Communication Technologies*,
Advances in Intelligent Systems and Computing 702,
https://doi.org/10.1007/978-981-13-0680-8_6

company wants to market a new product, it would like to identify a small target set of influential users that help in the spread of information through the effect of "word of mouth." Influence is the most important factor to form a social community in that network. Social communities are built from the group of nodes or members of the network who share a common interest in them. These communities consist of densely connected nodes to set of nodes but sparsely connected to other groups in the network [3]. The analysis of the evolution of communities is related to important social phenomena such as homophile [4] and influence [5]. Homophile is the phenomenon which resembles the proverb "birds of a feather flock together." In communities, users perform different interactive activities for exchanging of data. These activities are going to change the relations in the network which follows diploid scenarios. In the first scenario, people start new relationships in friend's network, which in turn change the structure of the network. As per second scenario, new director link may be added to the corporate network which changes data flow pattern. So these communities evolve/change over a time. In order to capture these changes, events and transitions of these communities have to be considered. The event of a community can be regarded as the evolution of a particular social community from one form to another form. It can be represented as a sequence of events (changes) such as form of a new community, dissolution of a community, survival of a community, bifurcation of a community, or merge of two or more communities, etc. [6]. Similarly, transition is defined as changes in features of the community, such as a change in the size, compactness, and persistence. These events and transitions are called life-cycle events. These life-cycle events cause different pattern change in these social communities. Finding patterns and analysis of these interactions have many applications, such as, in viral marketing [7], revenue maximization [8], and social influence [9]. The analysis of the evolution of communities gives the behavioral patterns of the communities such as type of interaction and detection of how the network evolves.

1.1 Background

There exists a significant amount of literature related to social network analysis. However, most of the existing techniques which focus on analysis of these models are limited to a single snapshot graphs. Graph is a representation of relation between nodes and edges of the social network. In social networking, in the evolution process, either new node will be added or new edge will be added. Most of the ongoing research focuses on collecting the data during a long period and then aggregating those data to create a large static graphical network. But real data are constantly evolving in nature [10]. Therefore, the new edges and nodes get masked due to the aggregation process, thereby resulting in loss of information. To address this type of issues, dynamic social networks are to be considered. Community detection is a fundamental problem in a dynamic social network analysis. The existing approaches for community detection are stochastic block model, latent factor model, clustering model [11], etc. These existing techniques have not taken into consideration node features and edge weights. These social network nodes interact among themselves, and changes are accordingly reflected in the social network structure. So in these social networks, we observe social interactions among communities. This interaction can affect the information flow and behavioral changes throughout the social networks. Most of the existing methods are not considering interconnection between evolving social networks and pattern change in these graphs [10]. Finding the interaction among the evolving social networks have many applications [12], such as evolution of unknown groups in an organization can influence into the organization's global decision-making behavior. It can help decision makers to set up profitable marketing strategies in advance. Social network community structures may undergo different events and transitions whenever evolution occurred in the social community. The occurrence of different events and transitions of the social community will be used for predicting the evolution of the community. Previous studies are not focused on predicting the events and transitions of the dynamic social communities.

One of the applications of SNA with the above-mentioned properties is "interlocking directors network." In our experiment, we are considering the following interlocking director's network. Interlocking directors is defined as the linkages among corporations created by individuals who sit on two or more corporate boards from [13]. Directors serving on the boards of powerful companies often sit on multiple boards. Companies fill the positions on their boards of directors with varying strategies, which results in a network of relationships between companies. Companies are linked by shared board members. And directors themselves are linked to each other by sitting on the different boards. In recent times, there are more incidents of recruiting diverse and experienced group of individuals who have a variety of skills and connections. The study of interlocking directors is useful to define the corporate community and to learn on the functioning of boards. In this scenario, the following challenges have to be addressed.

(a) Using temporal evolution of property of network, we capture the patterns of the interaction between directors and companies within a given time frame (say from 2003 to 2004). This shows how relations are formed in the director network and gives events of change in the network.

(b) Detecting the changes (events) in groups for the given time frames and identification of chains preceding the recent state of the group which provides fertile ground for building the predictive model.

Our model captures the roles of nodes in the graph and how they evolve over time with respect to interactions among the nodes. So we implement our model for identifying community patterns, events based on the behavioral changes, predicting future structural changes, and detecting temporal behavior transitions.

The rest of the paper is structured as follows. The literature survey is presented in Sect. 2. The methodology, preliminary experiments, and results are included in Sects. 3 and 4, respectively.

2 Literature Survey

This section discusses the research that has been done in the field of social network analysis. Our survey is on community detection, evolving communities, and events in evolving communities.

2.1 Community Detection

Community is defined [14] as a cohesive group of nodes that are connected more densely to each other than to the nodes in other communities. Cohesiveness depends upon various measures, such as sharing similar characteristics best known as homophile. Informally, we can say it as "feathers of same birds flock together." By this intuition of homophile, people can form communities. A group of people that have a particular characteristic in common can be formed as communities.

2.2 Evolving Communities

Communities evolve based on the appearance of new pages in the network, updating of the content constantly on the web, beginning of new relationships by people, etc. Thus, recent research is based on the evolving communities, which are temporal in nature. The temporal evolution of social networks has attracted many researchers. Leskovec et al. [15] studied the patterns of growth for large social networks based on the properties of huge networks based on the degree of distribution. Kempe's [16] work provides the properties of two real-world networks and then analysis on the evolution of structure in the social networks. However, in these cases, the properties on the graph level are studied on the level of communities but not on the node level. Takaffoli et al. [6] proposed a mathematical and computational framework that enables tracking the evolution of communities. They provided a heuristic technique that involves greedily matching detected communities at different snapshots. Falkowski [17] discussed the evolution of communities by applying clustering on a graph formed by all detected communities at different time points. The aforesaid research laid a path for evolving networks to change their structure whenever events occur in the communities. Event can be defined as change of a particular social community from one form to another form. It can be represented as a sequence of events (changes) such as form of a new community, dissolution of a community, survival of a community, bifurcation of a community, or merger of two or more communities, etc. [6].

2.3 Events in Evolving Communities

Research is going in the direction of identifying the critical events. These events are characterizing the dynamic social networks and their evolution. Palla et al. [5] applied clique percolation method (CPM) community mining on a social network formed by the communities and identified events, which are discovered at two consecutive snapshots. It resulted set of events pertained to set communities. Asur et al. [18] applied bit operations on consecutive snap shots of the static graphs. The result of this work is critical events between the identified communities. But this experiment is not covering all the events and its corresponding transactions. Asur [18] also identified events on communities and individuals by using an event-based framework. Greene [19] described each community by the help of series of events using a weighted bipartite matching to map communities and then characterized each community by the help of a series of events. Takaffoli [12] captured events and transitions by implementing an event-based framework. This framework captures the events and transitions of communities and individuals in a given time frame. All these frameworks are validated via the extraction of the topics for each community. Newman [20] has proposed number of methods to address community detection problem. These are categorized into top-down computer science approaches and bottom-up sociological approaches. A more detailed explanation and examples of each are described in [20]. The great insight for a dynamic social network is detecting the evolution of communities by monitoring when they form, dissolve, and reform.

Based on the extensive review, the proposed approach addresses the following issues:

- Community detection in dynamic social networks.
- Investigating interaction between evolving social networks and pattern change analysis in the graphs.

3 Methodology

In this section, we are specifying the methodology we followed for implementing "analysis on temporally evolving social networks." This task is divided into two modules.

- We find the individuals who are sitting together to undertake actions in their communities.
- We find the events and transitions on social communities which cause changes in the structure of social networks.

Data. We collected data from the Securities and Exchange Board of India (SEBI). SEBI is an Indian organization; it maintains listed companies and board of directors of these companies. We gathered few listed companies data from 2002 to 2012. These data maintain directors list, company information, and appointment date of the directors. That appointment date will be useful for temporal evolution.

These data are rearranged into two types of graphs for each year to incorporate the temporal feature. First graph, Fig. 1a, is called as the director–director graph and other one is company–company graph. In director–director graph, the nodes are directors and edges will be formed between nodes, if both the directors are working in the same company in the same year. Similarly, Fig. 1b is called as company–company graph, nodes are the companies, and if there is a common director between the companies (nodes), then edge will be formed between those company nodes. This type of graphs we constructed for each year from 2002 to 2012. Node numbers are the unique id's given to directors and companies.

The main focus of our work is to detect the events when evolutional change occurred in the director–director graph. We observed the following changes in the director–director graph evolution by applying connected components techniques. In our work, we observed interlocking nodes, the nodes which come together in each temporal graph are influencing for taking some decisions. In director–director graph, the interlocking directors are influencing to pull the known directors into company's director position. This observation can be interpreted as a change in network evolution which resembles adding the new director to director–director graph. And the second observation is that director edge got added in the company–company graph. When that director's edge got added at the same time, its corresponding network also got added to the director–director graph. This can be interpreted as an event which resembles adding new edge in company–company graph causes structural change in the director–director graph. This can be explained with the following Fig. 2.

The **mathematical model** for our social network is the adjacency matrices for directors and also for companies. For each year, we constructed director matrix, i.e., D_{ij}^t. If director D_i and director D_j working in company C in year t, then $D_{ij}^t = 1$ else $D_{ij}^t = 0$. D_{ij}^t is a square matrix of size $n \times n$, where "n" is the number of directors. Similarly, we constructed company matrix, i.e., C_{ij}^t. If director D is part for company C_i and company C_j in year t, then $C_{ij}^t = 1$ else $C_{ij}^t = 0$. C_{ij}^t is also a square matrix of size $c \times c$. The "c" is the number of companies. In the existing

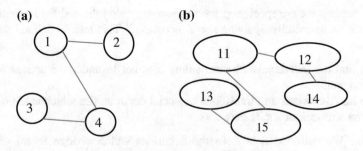

Fig. 1 **a** Company–company graph and **b** director–director graph

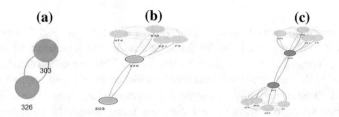

Fig. 2 a In year 2004, 303 and 326 are the dirctors of the the company Coal india. **b** we pointed out that the one of the directors of Coal india i.e., 326 had joined in the comapany Godawari. **c** represents graph of the year 2010. In 2010, the director of the company Coal India i.e., 326 had joined in NMDC. This scenario shows about new connections of the Directors

approaches, they constructed social network then the designed data structures, whereas in our approach, we extracted relation among the directors and companies and we put it in adjacency matrices. And we designed graphs. The advantage of our approach is we can easily find communities either row-wise or column-wise based on the behavioral changes of the social network. So the underlying behavior can be detected by following proposed algorithm:

Algorithm 1: Interlocking_Nodes
Step 1: Construct incidence matrix M, rows are directors and columns are companies (nXc).
If D_i is working in C_j then $M_{ij} = 1$ else $M_{ij} = 0$
Step 2: Compute $M * M^T$. It results a matrix of size $n \times n$. Diagonal elements are (D_{ii}) gives directors board membership. Non diagonal elements are board membership between D_i and D_j.
Step 3: Find the similarity among the directors in M using Jaccard similarity.

Algorithm 1 returns set of people sitting together, who cause structural change in the network. Jaccard similarity is a special kind of similarity index finding technique. It measures the overlap among the given entities. The maximum overlap entity value is 1, which means the participated elements come together in all locations. In our work, we wanted to find interlocking nodes, nothing but the overlapping nodes and also overlapping value must be 1. Hence, we applied Jaccard similarity on the incidence matrix M row-wise. As I mentioned earlier, our approach is suitable to find behavioral changes row-wise.

4 Results

In this experiment, we considered a data set of 575 directors and 149 companies. We used R language and R-studio for visualization and simulation of our implementation. These data we stored in R-data frame. Then we constructed adjacency matrix for each year. Next we calculated incidence matrix and applied our algorithm in R-environment only. We applied Jaccard similarity row-wise to the matrix M and then found out merge event directors using connected components techniques and used R's "igraph" for visualization. We calculated the *Influence proportion* using *Betweenness Centrality* measure (Table 1).

We are representing interlocking nodes and their influence in the following Fig. 3. It shows more influence proportion, if there are more interlocking directors. In the following Fig. 3, *x*-axis represents years and *y*-axis represents influence proportion.

Table 1 Proportion of merge event directors' influence

Year	No. of companies	No. of directors	No. of merge event directors	Proportion
2002	2	3	0	0
2003	21	38	0	0
2004	63	109	0	0
2005	70	125	0	0
2006	110	266	1	0.02
2007	129	340	1	0.03
2008	139	450	3	0.15
2009	142	527	8	0.15
2010	148	560	12	0.16
2011	149	566	13	0.16
2012	149	575	13	0.16

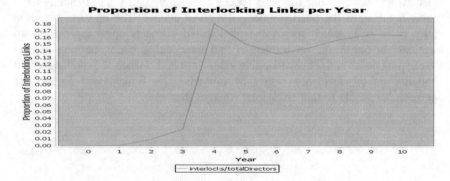

Fig. 3 Proportion of merge event directors

Table 2 List of directors working in more than one company

Year	Merge event directors
2002	0
2003	0
2004	0
2005	0
2006	310
2007	310
2008	310, 326, 453
2009	415, 310, 313, 326, 371, 453, 438, 527
2010	415, 310, 313, 494, 560, 303, 326, 371, 453, 438,527, 559
2011	415, 438, 310, 313, 494, 371, 453, 429, 560, 303, 326, 527, 559
2012	415, 438, 310, 313, 494, 371, 453, 429, 560, 303, 326, 527, 559

Following are the list of directors working in more than one company, which is year-wise data.

Table 2 shows year-wise interlocking directors list. The above calculation shows set of directors, who sit in more than one company. We need to find out set of directors sitting together in same set of companies in same year. These are called as interlocking directors, which can be able to find out by using above-mentioned algorithm. It results set of interlocking directors who cause structural change in the network. We calculated interlocking directors for every year and also how much proportion they are influencing.

Few of them are

- 310 and 453 are sitting together in oil corporation and Mangalore refinery since 2009.
- 527 and 559 are sitting together in NHPC and power grid since 2010.

5 Conclusion and Future Work

In this paper, we presented an algorithm for finding the events whenever there was a change in the structure of the network. That change has been identified by the Jaccard similarity. In our future work, we would like to find the structural features of these networks, which influence the temporal patterns of the networks. These structural features are very much essential to represent the relation among these temporal patterns. We also plan to apply predictive methodologies on these features to detect the future patterns of these temporal networks.

Acknowledgements The author is grateful to Securities and Exchange Board of India (SEBI) for providing the data set containing lists of Indian companies and other information required for the implementation of the algorithm.

References

1. Otte, E., Rousseau, R.: Social network analysis: a powerful strategy, also for the information sciences. J. Inf. Sci. **28**(6), 441–453 (2002)
2. Anagnostopoulos, A., Kumar, R., Mahdian, M. Influence and correlation in social networks. In: Proceedings of the 14th ACM SIGKDD International Conference on Knowledge Discovery and Data Mining, pp. 7–15. ACM, Aug 2008
3. Claudy, M.C., Garcia, R., O'Driscoll, A.: Consumer resistance to innovation—a behavioral reasoning perspective. J. Acad. Mark. Sci. **43**(4), 528–544 (2015)
4. Khurana, U., & Deshpande, A.: Efficient snapshot retrieval over historical graph data. In 2013 IEEE 29th International Conference on Data Engineering (ICDE), pp. 997–1008, Apr 2013
5. Palla, G., Barabási, A.L., Vicsek, T.: Quantifying social group evolution. Nature **446**(7136), 664 (2007)
6. Takaffoli, M., Sangi, F., Fagnan, J., Zäıane, O.R.: Community evolution mining in dynamic social networks. Procedia-Soc. Behav. Sci. **22**, 49–58 (2011)
7. Leskovec, J., Adamic, L.A., Huberman, B.A.: The dynamics of viral marketing. ACM Trans. Web (TWEB) **1**(1), 5 (2007)
8. Akhlaghpour, H., Ghodsi, M., Haghpanah, N., Mirrokni, V.S., Mahini, H., & Nikzad, A.: Optimal iterative pricing over social networks. In: International Workshop on Internet and Network Economics, pp. 415–423. Springer, Berlin, Dec 2010
9. Agarwal, N., Liu, H., Tang, L., Yu, P.S.: Identifying the influential bloggers in a community. In: Proceedings of the 2008 International Conference on Web Search and Data Mining, pp. 207–218. ACM, Feb 2008
10. Farajtabar, M., Wang, Y., Rodriguez, M.G., Li, S., Zha, H., Song, L.: Coevolve: a joint point process model for information diffusion and network co-evolution. In: Advances in Neural Information Processing Systems, pp. 1954–1962 (2015)
11. Zhang, Y., Levina, E., Zhu, J.: Community detection in networks with node features. Electron. J. Stat. **10**(2), 3153–3178 (2016)
12. Takaffoli, M., Zaïane, O.R.: Social network analysis and mining to support the assessment of on-line student participation. ACM SIGKDD Explor. Newsl **13**(2), 20–29 (2012)
13. Friel, N., Rastelli, R., Wyse, J., Raftery, A.E.: Interlocking directorates in Irish companies using a latent space model for bipartite networks. Proc. Natl. Acad. Sci. **113**(24), 6629–6634 (2016)
14. Sharma, A., & Cosley, D.: Distinguishing between personal preferences and social influence in online activity feeds. In: Proceedings of the 19th ACM Conference on Computer-Supported Cooperative Work & Social Computing, pp. 1091–1103. ACM, Feb 2016
15. Leskovec, J., McGlohon, M., Faloutsos, C., Glance, N., Hurst, M.: Patterns of cascading behavior in large blog graphs. In: Proceedings of the 2007 SIAM International Conference on Data Mining, pp. 551–556. Society for Industrial and Applied Mathematics, Apr 2007
16. Kempe, D., Kleinberg, J., Tardos, É.: Maximizing the spread of influence through a social network. In: Proceedings of the Ninth ACM SIGKDD International Conference on Knowledge Discovery and Data Mining, pp. 137–146. ACM, Aug 2003
17. Falkowski, T., Barth, A., Spiliopoulou, M.: Studying community dynamics with an incremental graph mining algorithm. In: AMCIS 2008 Proceedings, p. 29 (2008)
18. Asur, S., Parthasarathy, S., Ucar, D.: An event-based framework for characterizing the evolutionary behavior of interaction graphs. ACM Trans. Knowl. Discov. Data (TKDD) **3**(4), 16 (2009)
19. Greene, D., Doyle, D., Cunningham, P.: Tracking the evolution of communities in dynamic social networks. In: 2010 International Conference on Advances in Social Networks Analysis and Mining (ASONAM), pp. 176–183. IEEE, Aug, 2010
20. Boukrab, R.: CP decomposition for community identification in Networks. Bachelor's thesis, Universitat Politècnica de Catalunya (2017)

A Novel Construction Method of Intuitionistic Fuzzy Set from Fuzzy Set and Its Application in Multi-criteria Decision-Making Problem

Akanksha Singh, Dheeraj Kumar Joshi and Sanjay Kumar

Abstract In this paper, we mention and address the limitation of an existing conversion method of intuitionistic fuzzy set (IFS) from fuzzy set (FS) and propose a novel conversion method for IFS from fuzzy set. To compare the proposed conversion method, we apply it on a TOPSIS method of MCDM problem in which criteria weights are determined using intuitionistic fuzzy entropy. We take a real example in this study to compare the ranking of four organizations of different sectors.

Keywords Fuzzy set · Intuitionistic fuzzy sets · TOPSIS · MCDM

1 Introduction

Earlier probability theory was used in decision making problems to address the uncertainty which is caused by randomness. To handle the non-stochastic uncertainty, fuzzy set (FS) theory proposed by Zadeh [1] was introduced. FS theory is the extension of classical set theory and was integrated to handle inexact and imprecise data in multi-criteria decision making (MCDM) problem. In real-life decision making problems, sometimes decision maker finds fuzzy set not sufficient enough to represent the uncertainty. Even though fuzzy decision making methods handle the uncertainty, they fail to handle non-stochastic non-determinacy which occurs in the evaluation of decision making information. Atanassov [2, 3] introduced intuitionistic fuzzy set (IFS), which was widely used by various researchers in MCDM problems [4–10].

A. Singh · D. K. Joshi · S. Kumar (✉)
Department of Mathematics, Statistics and Computer Science, G. B. Pant University
of Agriculture and Technology, Pantnagar 263145, India
e-mail: skruhela@hotmail.com

A. Singh
e-mail: singhakanksha981993@gmail.com

D. K. Joshi
e-mail: maths.dj44010@gmail.com

© Springer Nature Singapore Pte Ltd. 2019
J. K. Mandal et al. (eds.), *Advanced Computing and Communication Technologies*,
Advances in Intelligent Systems and Computing 702,
https://doi.org/10.1007/978-981-13-0680-8_7

To handle non-determinacy of information, we need to change crisp data into IFS. In the literature, there is no direct method for the conversion of crisp sets into IFS. So firstly we have to construct FS from crisp set and then we construct IFS from FS. A construction method is developed by Jurio et al. [11] to construct IFS from FS. IFS, which is constructed using Jurio et al. [11] method, is widely used in time series forecasting [12–14]. In this paper, few limitations of Jurio et al. method [11] are discussed. We have also proposed a novel construction method of IFS from FS and its implementation in Technique for Order Preference by Similarity to Ideal Solution (TOPSIS) method.

2 Preliminaries

The definitions of FS, IFS, and conversion theorem [11] are presented in this section as follows:

Definition 1 [1] Let F be domain of discourse. A FS A_F is defined on F as follows:

$$A_F = \{(u, \mu_{A_F}(u)) | u \in F\} \tag{1}$$

here, $\mu_{A_F} : F \rightarrow [0, 1]$ is a mapping that defines membership grade of every element of F in FS A_F.

Definition 2 [3] An IFS A_I on F is defined as an object of following form:

$$A_I = \{\langle u, \mu_{A_I}(u), v_{A_I}(u) \rangle | u \in F\} \tag{2}$$

here, $\mu_{A_I} : F \rightarrow [0, 1]$ and $v_{A_I} : F \rightarrow [0, 1]$ are mappings that define membership grade and non-membership grade of every element of F in IFS A_I and, $\forall u \in F$, $0 \leq \mu_{A_I}(u) + v_{A_I}(u) \leq 1$ and is called degree of hesitation of IFS A_I in F with the following condition

$$\pi_{A_I}(u) = 1 - \mu_{A_I}(u) - v_{A_I}(u) \tag{3}$$

2.1 Construction of IFS from FS

The main characteristic of conversion method proposed by Jurio et al. [11] is that it fixes indeterminacy index for every element of F. The description of method is as follows:

Let $A_F \in \mathrm{F}_s(F)$ be set of FS on F. And let $\pi, \delta \colon F \to [0, 1]$ be two mappings, then

$$A_I = \left\{ \langle r_i, f(\mu_{A_I}(r_i), \pi(r_i), \delta(r_i)) \rangle \forall r_i \in F \right\} \tag{4}$$

is an IFS, where the mapping

$$f \colon [0, 1]^2 \times [0, 1] \to I^*$$
$$f(u, v, \delta) = (f_\mu(u, v, \delta), f_v(u, v, \delta))$$
$$f_\mu(u, v, \delta) = u(1 - \delta v) \text{ and } f_v(u, v, \delta) = 1 - u(1 - \delta v) - \delta v$$
$$\text{and } I^* = \{(u, v) : (u, v) \in [0, 1] \times [0, 1] \text{ and } u + v \leq 1\}$$

Satisfy the following conditions:

1. If $v_1 \leq v_2$, then $\pi(f(u, v_1, \delta)) \leq \pi(f(u, v_2, \delta))$ for all $u, \delta \in [0, 1]$.
2. $f_\mu(u, v_1, \delta) \leq u \leq 1 - f_v(u, v, \delta)$ for all $u \in [0, 1]$.
3. $f(u, 0, \delta) = (u, 1 - u)$.
4. $f(0, v, \delta) = (0, 1 - \delta v)$.
5. $f(u, v, 0) = (u, 1 - u)$.
6. $\pi(f(u, v, \delta)) = \delta v$.

Example 1 Let $F = \{r_1, r_2, r_3, r_4\}$ be a reference set, and let A_F be a FS given as follows:

$$A_F = \{(r_1, 0.2), (r_2, 0.8), (r_3, 1), (r_4, 0.5)\}$$

and $\pi(r_i) = 0.2\delta(r_i) = 1$ for all $r_i \in F$. By using methodology given in Sect. 2.1, we obtain the following IFS:

$$\mathbf{A}_F = \{(r_1, 0.16, 0.64), (r_2, 0.64, 0.16), (r_3, 0.80, 0), (r_4, 0.40, 0.40)\}$$

2.2 Limitations of the Existing Conversion Method Proposed by Jurio et al. [11]

1. In this method, indeterminacy index of every element of domain of discourse is fixed beforehand. This means degree of hesitancy for every element in a set is the same.
2. In this method, degree of membership grade of an element is impractically changed when IFSs are constructed from FS. As in Example 1, membership grade of element r_1 is 0.2, and when IFS is constructed using this method, the membership is changed into 0.16.

3. This method includes indeterminacy forcefully even in those elements which have no uncertainty. As in Example 1, membership grade of element r_3 is 1, i.e., it has no fuzziness, but when IFS is constructed using method given by Jurio et al. [11], the membership grade is changed into 0.80 and raises the uncertainty of 0.20 in r_3.

The following example illustrates the limitations of Jurio et al. [11] method more efficiently.

Example 2 Let us consider an example of voting system to understand aforesaid limitations. Let E be set of countries with elective governments. Let $M(u)$ denote percentage of electorate voted for the corresponding government of country $u \in E$ and $\mu(u) = \frac{M(u)}{100}$.

This number corresponds to that part of electorate who has voted for the government named as membership grade $\mu(u)$ of u. The number $1 - \mu(u)$ corresponds to that part of electorate who has not voted for government. To consider the value $1 - \mu(u)$ in more detail, we need IFS where we define $v(u)$ as the number of votes given outside the government called non-membership grade. For the part of electorate who has not voted at all called degree of uncertainty $\pi(u) = 1 - \mu(u) - v(u)$.

3 Novel Method to Construct IFS from FS

This section presents novel method to construct IFS from FS. In FS, there is only one parameter known as membership grade of elements $\mu_{A_F}(u)$. But in IFS, there are three parameters: membership grade, non-membership grade, and degree of hesitancy. It is more convincing to modify non-membership grade of an element to include non-determinacy rather than membership grade. Therefore, the membership grade of element should not change when IFS is constructed from FS. In other words, the appearance of non-membership and uncertainty should not affect membership grade of element. The novel construction method is as follows:

Proposition *The mapping F: $[0, 1] \rightarrow [0, 1] \times [0, 1]$ is given by*

$$F(u) = (F_\mu(u), F_v(u)) \tag{5}$$

$$\text{here, } F_\mu(u) = u \quad \text{and} \quad F_v(u) = g(1 - u) \tag{6}$$

here, g: $[0, 1] \rightarrow [0, 1]$ is any mapping such that

$$g(0) = 0 \text{ and } g(u) \leq u \tag{7}$$

Theorem *Let A_F be any FS, then $A_F = \{(u, \mu_{A_F}(u)) | u \in F\}$ and g: $[0, 1] \rightarrow [0, 1]$ be any mapping. Then,*

$$A_I = \{(u, F(\mu_{A_F}(u)))|u \in F\} \tag{8}$$

is an IFS.

i.e., $A_I = \{(u, F_\mu(\mu_{A_F}(u), F_v(v_{A_F}(u))|u \in F\}$ *with*

$$F_\mu(\mu_{A_F}(u)) = \mu_{A_F}(u) \quad and \quad F_v(\mu_{A_F}(u)) = g(1 - \mu_{A_F}(u)) \tag{9}$$

Here, g can be defined to determine non-membership or to determine indeterminacy of IFS, having property

$$g(0) = 0 \quad and \quad g(u) \leq u \tag{10}$$

Hence, g is called non-membership function or indeterminacy function according to its purpose of defining.

Proof Proof of the theorem immediately follows from Eq. 5 with $u = \mu_{A_F}(u)$

Corollary 1 *If in the theorem,*
$\mu_{A_F}(u) = 1$ *for any* $u \in F$, *then* $F_v(\mu_{A_F}(u)) = 0$ *and* $\pi_{A_F}(u) = 0$.

Proof Proof of this corollary immediately follows from Eq. 9 with $\mu_{A_F}(u) = 1$, and using Eq. 10, $F_v(\mu_{A_F}(u)) = 0$, after that using Eq. 3, we will get $\pi_{A_F}(u) = 0$.

Corollary 2 *If* $g(1 - \mu_{A_F}(u)) = c$ *and* $\mu_{A_F}(u) \neq 1$ *here,* $c \in [0, 1]$, *is fixed indeterminacy index for IFS* A_I. *Then,* $\pi_{A_F}(u) = c \forall u \in F$

Proof If we take $g(u)$ as indeterminacy function which is defined by $g(u) = c$, then for any $u \in F$ indeterminacy is fixed to 'c' and $\pi_{A_F}(u) = c \forall u \in F$. This proves the Corollary 2.

The following example illustrates the proposed conversion method from FS to IFS.

Example 3 Let $F = \{r_1, r_2, r_3, r_4\}$, and let A_F be a FS, given by

$$A_F = \{(r_1, 0.2), (r_2, 0.8), (r_3, 1), (r_4, 0.5)\}$$

On defining non-membership function $g: [0, 1] \to [0, 1]$ such that:

$$g(u) = \frac{u}{a},$$

here, 'a' is sum of all u.

Using above theorem, obtain the following IFS:

$$A_I = \{(r_1, 0.2, 0.53), (r_2, 0.8, 0.13), (r_3, 1.0, 0.0), (r_4, 0.5, 0.16)\}$$

In order to compare the constructed IFSs in Example 1 and Example 3, we calculate intuitionistic fuzzy entropy using the following expression:

$$E(I) = \frac{1}{n} \sum_{i=1}^{n} \frac{\min(\mu_I(r_i), v_I(r_i)) + \pi(r_i)}{\max(\mu_I(r_i), v_I(r_i)) + \pi(r_i)}$$

As $E(\mathbf{A}_F) = 0.514$ and $E(\mathbf{A}_I) = 0.353$, it confirms that the proposed conversion method constructs more accurate IFS than that of Jurio et al. [11].

4 Implementation of the Proposed Conversion Method in Intuitionistic Fuzzy TOPSIS

TOPSIS [15] is one of the famous MCDM methods. TOPSIS is based on the concept of maximum and minimum distance of the best option from positive ideal solution (PIS) and negative ideal solution (NIS), respectively. This method was extended to fuzzy TOPSIS by Chen [16].

In this section, an example [4] is taken to the implementation of the proposed construction method. In this example, algorithm [4] of intuitionistic fuzzy TOPSIS is implemented to rank four companies: Bajaj Steel (O_1), HDFC. Bank (O_2), Tata Steel (O_3), and Infotech Enterprises (O_4) using the following five criteria (Cr_1, Cr_2, Cr_3, Cr_4, Cr_5):

1. Earnings per share (EPS) of company (Cr_1),
2. Face value (Cr_2),
3. Put/call ratio (P/C ratio) of company (Cr_3),
4. Dividend yield of company (Cr_4), and
5. Price-to-earnings ratio (P/E ratio) of company (Cr_5).

Criteria Cr_1 and Cr_2 belong to benefit criteria; i.e., growth prospects are reflected by their high value. Criteria Cr_3, Cr_4, Cr_5 belong to non-beneficial criteria; i.e., growth prospects are reflected by their low value. Table 1 [4] shows the average values of numerical data.

Step 1. Numerical values of criteria (Table 1) are fuzzified. Fuzzy decision matrix is shown in Table 2.

The proposed construction method (Sect. 3) is applied on fuzzy decision matrix (Table 2) to construct the following IF decision matrix (Table 3).

Table 1 Numerical values of criteria

	Cr_1	Cr_2	Cr_3	Cr_4	Cr_5
O_1	20.50	10	2.17	2.08	4.69
O_2	23.31	2	24.3	0.69	27.73
O_3	60.06	10	5.63	2.96	6.60
O_4	16.86	5	9.5	1.28	11.59

Table 2 Fuzzy decision matrix

	Cr_1	Cr_2	Cr_3	Cr_4	Cr_5
O_1	0.2846	0.75	0.165	0.56	0.24
O_2	0.318	0.248	0.74	0.232	0.76
O_3	0.759	0.75	0.26	0.77	0.149
O_4	0.241	0.437	0.357	0.37	0.39

Table 3 Intuitionistic fuzzy decision matrix

	Cr_1	Cr_2	Cr_3	Cr_4	Cr_5
O_1	(0.284, 0.298)	(0.75, 0.138)	(0.165, 0.337)	(0.56, 0.213)	(0.24, 0.309)
O_2	(0.318, 0.284)	(0.248, 0.414)	(0.74, 0.105)	(0.232, 0.371)	(0.76, 0.098)
O_3	(0.759, 0.101)	(0.75, 0.138)	(0.26, 0.299)	(0.77, 0.111)	(0.149, 0.346)
O_4	(0.241, 0.317)	(0.437, 0.31)	(0.357, 0.259)	(0.37, 0.305)	(0.39, 0.248)

Step 2. Criterion weights are determined using IF entropy [17] as follows:

$$W = \{0.225, 0.173, 0.214, 0.178, 0.211\}$$

Step 3. IF decision matrix (Table 3) is multiplied by the weights of criterion to have the following weighted IF decision matrix (Table 4).

Step 4. IFPIS and IFNIS are calculated [8] and are as follows:

$$A^+ = \{(0.274, 0.597), (0.213, 0.71), (0.038, 0.792), (0.046, 0.838), (0.033, 0.799)\}$$
$$A^- = \{(0.06, 0.772), (0.048, 0.859), (0.25, 0.792), (0.23, 0.838), (0.26, 0.799)\}$$

Step 5. Distance of each alternative from IFPIS and IFNIS is calculated and is given in Table 5.

Step 6. According to decreasing order of closeness coefficient (Cc) values, preference sequence is obtained as $O_3 > O_1 > O_4 > O_2$. Hence, among four companies (O_1, O_2, O_3, O_4), the best company is O_3.

Table 4 Weighted intuitionistic fuzzy decision matrix

	Cr_1	Cr_2	Cr_3	Cr_4	Cr_5
O_1	(0.072, 0.762)	(0.213, 0.71)	(0.038, 0.792)	(0.136, 0.759)	(0.056, 0.781)
O_2	(0.083, 0.753)	(0.048, 0.859)	(0.25, 0.617)	(0.046, 0.838)	(0.26, 0.613)
O_3	(0.274, 0.597)	(0.213, 0.71)	(0.062, 0.772)	(0.23, 0.676)	(0.033, 0.799)
O_4	(0.06, 0.772)	(0.095, 0.817)	(0.095, 0.817)	(0.079, 0.809)	(0.099, 0.745)

Table 5 Intuitionistic separation measures

Alternatives	S_i^+	S_i^-	Closeness coefficient (Cc)
O_1	0.315	0.687	0.686
O_2	0.795	0.568	0.417
O_3	0.208	0.956	0.821
O_4	0.483	0.519	0.518

Table 6 Preference order of alternatives using different TOPSIS methods

Method	Ranking	Best alternative
Proposed by Grzegorzewski and Mrówka [18]	$O_3 > O_4 > O_1 > O_2$	O_3
Proposed by Hung and Chen [19]	$O_3 > O_1 > O_4 > O_2$	O_3
Proposed by Joshi and Kumar [4]	$O_3 > O_1 > O_4 > O_2$	O_3
Proposed method	$O_3 > O_1 > O_4 > O_2$	O_3

5 Comparative Analysis

In order to validate and compare developed method, the preference order of these four alternatives is also obtained by using TOPSIS and fuzzy TOPSIS method developed by Hung and Chen [19]. Table 6 shows the ranking order using different methods.

It is clear from Table 6 that there is no confliction in the ranking order of best and worst alternatives by fuzzy TOPSIS, intuitionistic fuzzy TOPSIS, and the proposed method.

6 Conclusion

In this paper, few limitations of construction method proposed by Jurio et al. [11] are mentioned and a novel construction method is proposed to address those limitations. IF entropy is applied to verify the improvement in constructed IFS using the proposed conversion method. We use this construction method to propose IFS-based TOPSIS. To compare the proposed conversion method, we apply it on a real example [4] to rank four enterprises using five inter-independent criteria. As no change in preference order is found using IF TOPSIS with the proposed and Jurio et al. [11] construction method, it confirms that the proposed construction method is valid.

References

1. Zadeh, L.A.: Fuzzy sets. Inf. Cont. **8**(3), 338–353 (1965)
2. Atanassov, K.T.: Intuitionistic fuzzy sets. Fuzzy sets Syst. **20**(1), 87–96 (1986)
3. Atanassov, K.T.: Intuitionistic fuzzy sets. Intui. Fuzzy Sets, 1–137 (1999) (Physica-Verlag HD)
4. Joshi, D., Kumar, S.: Intuitionistic fuzzy entropy and distance measure based TOPSIS method for multi-criteria decision making. Egypt. Info. **15**(2), 97–104 (2014)
5. Szmidt, E., Kacprzyk, J.: Intuitionistic fuzzy sets in group decision making. Notes IFS **2**(1) (1996)
6. Szmidt, E., Kacprzyk, J.: Using intuitionistic fuzzy sets in group decision making. Cont. Cyber. **31**, 1055–1057 (2002)
7. Pankowska, A., Wygralak, M.: General IF-sets with triangular norms and their applications to group decision making. Inf. Sci. **176**(18), 2713–2754 (2002)
8. Wan, S.P., Li, D.F.: Atanassov's intuitionistic fuzzy programming method for heterogeneous multiattribute group decision making with Atanassov's intuitionistic fuzzy truth degrees. IEEE Trans. Fuzzy Syst. **22**(2), 300–312 (2014)
9. Wan, S.P., Yi, Z.H.: Power average of trapezoidal intuitionistic fuzzy numbers using strict t-norms and t-conorms. IEEE Trans. Fuzzy Syst. **22**(2), 300–312 (2015)
10. Xian, S., Xue, W., Dong, Y.: Intuitionistic fuzzy induced ordered entropic weighted averaging operator for group decision making. J. Int. Fuzzy Syst. **31**(3), 1189–1197 (2016)
11. Jurio, A., Paternain, D., Bustince, H., Guerra, C., Beliakov, G.: A construction method of Atanassov's intuitionistic fuzzy sets for image processing. In: 5th IEEE Conference on Intelligent System, London, UK (2010)
12. Joshi, B.P., Kumar, S.: Intuitionistic fuzzy sets based method for fuzzy time series forecasting. Cyber. Syst. **43**(1), 34–47 (2012)
13. Joshi, B.P., Kumar, S.: Fuzzy time series model based on intuitionistic fuzzy sets for empirical research in stock market. Inter. J. App. Evolut. Comput. (IJAEC) **3**(4), 71–84 (2012)
14. Gangwar, S.S., Kumar, S.: Probabilistic and intuitionistic fuzzy sets-based method for fuzzy time series forecasting. Cyber. Syst. **45**(4), 349–361 (2014)
15. Hwang, C.L., Yoon, K.S.: Multiple Attribute Decision Making Methods and Applications. Springer, Berlin (1981)
16. Chen, C.T.: Extension of the TOPSIS for group decision making under fuzzy Environment. J. Fuzzy Sets Syst. **114**, 1–9 (2000)
17. Szmidt, E., Kacprzyk, J.: Entropy for intuitionistic fuzzy sets. Fuzzy Sets Syst. **118**(3), 467–477 (2001)
18. Grzegorzewski, P., Mrówka, E.: Some notes on (Atanassov's) intuitionistic fuzzy sets. Fuzzy Sets Syst. **156**(3), 492–495 (2005)
19. Hung, C.C., Chen, L.H.: A multiple criteria group decision making model with entropy weight in an intuitionistic fuzzy environment. Int. Auto. Com. Eng., pp. 17–26 (2009) (Springer, Netherlands)

Design of an Audio Repository for Blind and Visually Impaired: A Case Study

**Ritu Singhal, Archana Singhal, Mahima Bhatnagar
and Nishita Malhotra**

Abstract The emerging cloud computing which enables ubiquitous access to shared pools of configurable system resource has boosted the higher education system by minimizing the IT infrastructure cost with improved manageability and less maintenance. However, issues related to modeling, designing, and evaluating interfaces between human and computing system for blind and visually impaired, leveraging cloud computing are yet to be fully explored. The design of an easy-to-use interface for visually impaired students is outlined in this paper, thereby exemplifying how advancements in technology could remove barriers of equal access to information and service. The objective of the proposed android-based design is to create an audio file sharing platform which involves creating a repository of the frequently used lectures, reading materials for these students. A case study is presented, illustrating the utility of proposed design. The proposed model allows the visually impaired students to access the audio repository both in online and offline modes from cloud storage. Benefits of the proposed design, based on the service models of cloud computing, are efficiency, scalability, fault tolerance, and 24 × 7 accessibility.

Keywords Cloud computing · SaaS · Android application · Visually impaired
Google Drive · Cloud storage · Audio repository

1 Introduction

Cloud computing paradigm (CCP) enables "ubiquitous, convenient, on-demand network access to a shared pool of configurable computing resources" [1]. Essential characteristics of CCP are on-demand service, broad network access, resource pooling, rapid elasticity, and measured service. The corresponding service models are Infrastructure as a Service (IaaS), Platform as a Service (PaaS), Software as a

R. Singhal (✉) · A. Singhal · M. Bhatnagar · N. Malhotra
Indraprastha College for Women, University of Delhi, New Delhi 110054, India
e-mail: toritu7@gmail.com

© Springer Nature Singapore Pte Ltd. 2019
J. K. Mandal et al. (eds.), *Advanced Computing and Communication Technologies*,
Advances in Intelligent Systems and Computing 702,
https://doi.org/10.1007/978-981-13-0680-8_8

Service (SaaS), and the deployment models are private, community, public, and hybrid. Cloud computing thereby offers infrastructural facility including data storage space, and applications online. Cloud computing providing latest and configurable resources in cost-effective manner has mitigated infrastructural needs for the academic sector and/or small and medium enterprises with limited budget. Cloud storage [2], a highly virtualized infrastructure, is a very useful service which allows off-site storage of data managed by the service providers. The archived data is accessible 24×7 by the applications. Service providers, e.g., Amazon, Google, and Microsoft, have built data center facilities around regions and availability zones with low latency, high throughput, and highly redundant networking, offering customers an easier and more effective way to design and operate applications and databases. Salient aspects of cloud computing are elasticity, pay per usage model, reduced requirement of expertise, reduced maintenance, reduced in-house infrastructure liability, and cost [1].

Human–computer interface, on the other hand, is concerned with issues related to modeling, designing, and evaluating interactions between humans and computing systems. With the advancements in technology, academic systems are rapidly changing, leading to renovation of the structures within the educational system as well as on the methodologies of teaching-learning. However, blind interaction and integration of visually impaired users with normal users are the key issues and challenges, hindering equal access to information and service.

Students in higher education, nowadays, require exposure to latest technology for survival in the competitive world. For students with visual disabilities (SVD), it becomes even more difficult to explore and access the study resources. It is the general practice among the SVD to record lectures and get their assignments written by fellow students. Books, articles, and other reading material are also recorded with the help of other students of the college. Moreover, after completing the degrees, the recorded study materials are rarely preserved. It is also very difficult for them to index or manage the old study material without any support. Library provides the study material according to their demands, but it is difficult for the students to keep them in proper order as per the syllabus or their area of interest. It has been seen that there is no common facility for keeping these books, recordings and sharing them among students with visual disabilities.

With the advancement of technology, various specialized devices like Lex Air camera, document scanner, Braille books, Braille printer, Daisy books, and software have helped in removing some of these academic barriers.

Information technology, in recent times has facilitated online lectures series, virtual classrooms, and distance education for remote students over Internet. However, these are inaccessible to SVD. The theme of this paper is concerned with how cloud computing technology could facilitate visually challenged students for accessing educational resources at the college or university level according to their requirements, removing the physical barriers. The paper proposes a generic architecture that represents a better coordination and collaboration among faculty and visually challenged students. It can be customized as per the needs of different educational institutes.

The remaining of the paper is structured as follows: Section 2 presents the literature review and the related works. Section 3 presents a brief background for understanding the design. Architecture of the proposed design is outlined in Sect. 4. Section 5 exemplifies the design with a case study, and in Sect. 6, conclusions are drawn.

2 Literature Review

Ashraf and Raza elaborated the problems of SVD due to lack of technological advancements [3]. In [4], Bocconni et al. discussed the issues related to the accessibility of educational material by visually impaired students. In [5], Şimşek et al. discussed various problems encountered by visually impaired students while developing information and communication technologies (ICT) skills, and suggested some solutions for these problems. In [6], Carolina v. Puffelen focused on the relationship between the ICT-related training offered to visually challenged people and their actual competencies for online activities. A wide range of assistive devices are demonstrated for SVD in [7]. Grussenmeyer and Folmer have recently reviewed touch screen accessibility for visually impaired persons [8].

Singhal et al. [9] proposed a cloud architecture that could benefit conventional education system with cloud resources, consequently enhancing the infrastructural needs of educational institutes with limited budget and physical resources. Many specialized applications and assistive technologies are now available for smartphones or tablets compatible with screen reading or screen magnification software including optical character recognition, text-to-speech features, GPS, and route finding [10].

Pacheo et al. [11] observed, the vision impaired students feel that ICT not only facilitate them in learning, information collection and communication, compensating their impairments, but also provide the chances to collaborate, social connection, support arrangement, and participation. Sahasrabudhe and Palvia [12] reported a qualitative and subsequent quantitative inquiry to understand the academic challenges faced by SVD, their resolution strategies, and the use of technology to resolve their problems with reference to India. They advocated that solutions using technology need to be developed in areas that do not exist [12].

In the present work, an android-based mobile application has been developed. The application involves creating a cloud-based audio repository for the readings and lectures, based on the prescribed curriculum. This repository is accessible through a mobile-based android application called "Lecture Hall."

3   Background

3.1   Available Technology

IT infrastructure can be used in both learning and communication by SVD. Choice of correct equipment and software depends on user's visual needs, and familiarity with the device. It includes usage of various software available for differently abled students like Daisy [13], JAWS [14], and Lex Air [15]. It has been observed that SVD can effectively use the above-mentioned software with minimal training. In the next section, the interface, i.e., voice-assistant features that enable blind people to use smartphone, is elaborated.

3.2   Study of Interface for Visually Challenged Students

Mobile accessibility is gaining popularity among SVD due to voice-assistant [16] feature. TalkBack is a voice-assistant feature of Google's Android Accessibility Service. It is designed to help people to listen to text instead of reading with their mobile devices. TalkBack application reads the text aloud written on the screen. Every movement of user on the phone is monitored and spoken. Thus, the visually impaired people can use a common android smartphone equipped with specific applications effectively for connectivity and online accessibility. During a preliminary survey, differently abled students have shown great interest and eagerness in using such mobile applications ensuring adequate usage of the application.

3.3   Classification of Course Structure in a University

Every university has varying course structure including core papers, elective, minor, or skill-based papers. In Delhi University, a choice-based credit system (CBCS), a UGC initiative to bring equity, efficiency, and excellence in the higher education system, is followed. Hierarchy in different semesters includes core course, elective course, discipline-specific elective papers (DSE), generic elective (GE), ability enhancement compulsory courses (AECC) and skill development courses/foundation course including environmental science, English communication/MIL communication, and AE elective course (AEEC) for providing value-based or skill-based instruction. Indexing can help students to navigate conveniently to required combination of papers coming from different disciplines.

3.4 Survey

A preliminary survey was conducted to find out the usability of an android-based application for visually challenged students. The methodology included in-person interviews and questionnaire to gather feedback on the current facilities and to find out effectiveness of such an application. As per the survey, majority of the students opined that a centralized facility of recorded lectures would be advantageous and they were comfortable in using their smartphones in accessing educational resources, using the voice-assistant facility TalkBack. This was the motivation behind the proposed design outlined in the next section.

4 Proposed Design

The aim of the design was to use the emerging cloud computing model for accessing resources according to university guidelines, anywhere at any time by the SVD. A common android platform "Lecture Hall" is designed for these students for using recorded reading materials and audio lectures as per the university syllabus. Data retrieval process is designed after studying the university-defined structure of courses, so that students can access these resources without any complex formalities. This application includes audio study material in both Hindi and English medium.

The proposed application model allows both offline and online access to the repository of their course content on their phone or tablet.

The five-layered architecture of the proposed design is shown in Fig. 1. These layers are as follows:

(i) **Storage layer**: The lowest layer is storage layer. It represents the cloud storage of the study material, readings, and local storage in user's device. In this layer, data is accessed by authorized users only.

(ii) **Service layer (SaaS)**: Second layer is the service layer. It consists of services offered to students with visual impairment. It includes the application created according to university-defined course structure. "Lecture Hall" is part of this layer.

(iii) **Cloud management layer**: Management layer deals with user's request. Services are provided to authorized users. Only read access is permitted for students. Additional access for updating the data is provided faculty.

(iv) **Authentication layer**: The role of this layer is to verify the authentic users for authorization purpose. Different levels of access rights are given to different users in this layer. Access is not given to unauthorized user.

(v) **Users layer**: Users in this layer include authorized users like faculty, visually impaired students, and administrators.

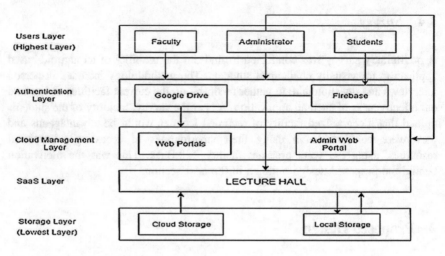

Fig. 1 Proposed layered architecture

The application is developed in two modules. The first module, i.e., the front-end development, includes the designing of an easy-to-use interface. The second module, i.e., back-end module, focuses on developing an audio repository of required material. To speed up the application development, both the modules were developed in parallel. Recorded study material is uploaded in cloud storage and linked to the user interface. Firebase is used for authentication and for real-time database management in the design. It is the mobile back-end service that provides three features: (i) user authentication, (ii) real-time database, and (iii) hosting. Flow diagram of the application "Lecture Hall" is shown in Fig. 2.

In the next section, a case study has been presented to discuss the functioning of the proposed approach.

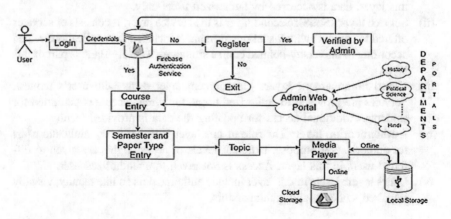

Fig. 2 Flow diagram of "Lecture Hall"

5   Case Study

Enabling Unit (EU) and Equal Opportunities Cell of IP College, Delhi, is dedicated toward taking care of the needs of students with disabilities in both undergraduate and postgraduate programs. The objective is to create barrier-free environment and provide accessible means for SVD. EU of the college has facilities like Lex Air camera, embosser, Daisy books, document scanner, Braille books, computers equipped with JAWS. As a consequence, the college received the 6th NCPEDP-Mphasis (National Centre for Promotion of Employment for Disabled People and Mphasis) Universal Design Awards 2015 under Category C (Companies/Organizations) for "creation of an environment where students with disabilities are brought into the mainstream of institutional life" [17].

Amazon, dropbox, firebase, and Google Drive cloud storages were explored and analyzed for providing a cost-effective solution. Google Drive was selected because of its comparative flexibility. It is a file storage and synchronization service, and offers 15 GB of free storage and maximum of 30 TB through optional payment plans. Files uploaded can be of maximum 5 TB in size. The screenshots of the developed application are depicted in Figs. 3 and 4. Figure 3a displays the first screen when the application is launched which asks for the sign in option and move to the option either to sign out or to continue. If user clicks on Lecture Hall button to continue, then the second screen shown in Fig. 3b appears with the options for usage of offline or online repository. After clicking top right corner, this screen displays the different options about the application like About Application, Disclaimer, Acknowledgement, Feedback, and Contact Us. Next screen displayed in Fig. 3c asks for the selection of department.

Appropriate semester is chosen afterward as shown in Fig. 4. After selecting the semester, one has options for the choice of courses followed by subject category (like Core, GEC, AECC). Then, students have to choose desired subject category

Fig. 3 Welcome and department screens of "Lecture Hall"

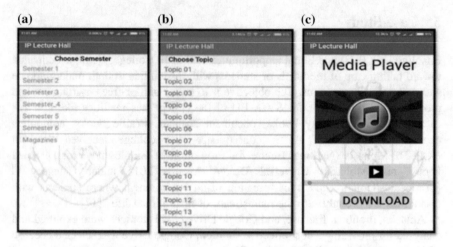

Fig. 4 Screens for, semester, topics and media player option

and various subject options under particular subject category/paper titles, followed by the topics available under particular subject option.

In Fig. 4c, the media player with the options to either play or download the files for offline use is displayed. The files will be saved in the "Downloads" folder automatically and can be accessed in offline mode.

6 Conclusion and Future Work

Integrations of visually impaired users with normal users are the key issues and challenges for equal access to information and services in the twenty-first century, "to remove the barriers to and in learning, to realize the full and equal participation of all persons with disabilities in society" [18]. The problem is more acute in India, where there were 7.2 million blind people in 1990 that rose to 8.8 million in 2015, making the country the home of almost a quarter of the total 36 million blind people; according to a study by Lancet Global Health Journal in 2015 [19]. This paper demonstrated how the emerging cloud computing technology can improve accessibility to educational resources by SVD. Almost 95% students, currently using the application, found it indispensable for examination preparation as well as survival in a knowledge-driven world, thereby validating utility of the proposed design. It can be argued in conclusion that the android application is an effective cognitive assistant tool for the SVD.

Acknowledgements The authors are grateful to Dr. Babli Moitra Saraf, IP College, University of Delhi, for valuable discussions during the design and implementation phases of the project. The proposed "Lecture Hall" has been designed as a part of innovation project under "Centenary Grant for Under Graduate research" sanctioned by the IP College, Delhi.

References

1. Mell, P., Grance, T.: The NIST Definition of Cloud Computing, NIST Special Publication 800-145, National Institute of Standards and Technology Gaithersburg, MD 20899-8930, pp. 1–7 (2011)
2. Rhea, S., Wells, C., Eaton, P., Geels, D., Zhao, B., Weatherspoon, H., Kubiatowicz, J.: Maintenance-free global data storage. IEEE Int. Comput. **5**(5), 40–49 (2001)
3. Ashraf, A., Raza, A.: Usability issues of smart phone applications: for visually challenged people. World Acad. Sci. Eng. Technol. Int. J. Comput. Electr. Autom. Control Inf. Eng. **8**(5), 760–767 (2014)
4. Bocconi, S., Dini, S., Ferlino, L., Martinoli, C., Ott, M.: ICT educational tools and visually impaired students: different answers to different accessibility needs. In: Stephanidis, C. (eds) Universal Access in Human-Computer Interaction. Applications and Services. UAHCI 2007. Lecture Notes in Computer Science, vol. 4556. Springer, Berlin (2007)
5. Şimşek, Ö., Altun, E., Ateş, A.: Developing ICT skills of visually impaired learners. Procedia-Soc. Behav. Sci. **2**(2), 4655–4661 (2010)
6. Van Puffelen, C.: ICT-related skills and needs of blind and visually impaired people. ACM SIGACCESS Access. Comput. **93**, 44–48 (2009)
7. https://www.teachingvisuallyimpaired.com/overview-of-assistive-technology.html
8. Grussenmeyer, W., Folmer, E.: Accessible touchscreen technology for people who are blind: a survey. ACM Trans. Access. Comput. http://dx.doi.org/ (2017)
9. Singhal, R., Singhal, A., Sonia.: Towards a generic E-Cloud architecture for universities. Int. J. Web Appl. **8**(2), 36–43 (2016)
10. https://www.afb.org/ProdBrowseCatResults.asp?CatID=102. Copyright © 2018 American Foundation for the Blind
11. Pacheco, E., Lips, M., Yoong, P.: Transition 2.0: digital technologies, higher education, and vision impairment. Internet Higher Educ. **37**, 1–10 (2018)
12. Sahasrabudhe, S., Palvia, P.: Academic challenges of blind students and their mitigation strategies. In: Proceedings of the Nineteenth Americas Conference on Information Systems, Chicago, Illinois, 15–17 Aug 2013
13. Kerscher, G.: DAISY consortium: information technology for the world's blind and print-disabled population-past, present, and into the future. Libr. HiTech19 **1**, 11–15 (2001)
14. http://www.freedomscientific.com/Products/Blindness/JAWS
15. http://www.visionaid.com/phpincludes/en/products/lex_air/lex_air_desc.php.   VisionAid International Ltd. Website
16. https://support.google.com/accessibility/android/answer/6006589?hl=en,©2018
17. http://www.ncpedp.org/sites/all/themes/marinelli/documents/NCPEDP-UD-Awards-2015-Web-Brochure.pdf
18. The Right to Education for Persons with Disabilities, United Nations Educational, Scientific and Cultural Organization 7, place de Fontenoy, 75352 Paris 07 SP, France (2015). http://unesdoc.unesco.org/images/0023/002325/232592e.pdf
19. Bourne, R.R.A., Resnikoff, S., Ackland, P., Braithwaite, T., Cicinelli, M.V., Das, A., Zheng, Y.: Global causes of blindness and distance vision impairment 1990–2020: a systematic review and meta-analysis vision loss expert group of the global burden of disease study. Lancet Global Health **5**(12), e1221–e1234 (2017)

A Futuristic Deep Learning Framework Approach for Land Use-Land Cover Classification Using Remote Sensing Imagery

Rahul Nijhawan, Deepankar Joshi, Naman Narang, Aditya Mittal and Ankush Mittal

Abstract Our aim is to propose a new deep learning framework approach which uses an ensemble of convolutional neural network (CNN) for land use-land cover mapping. Every CNN layer was fed with diverse combination of multispectral and geospatial satellite bands provided by Sentinel 2 satellite imagery (spatial resolution of 10 m), topographic and derived texture parameters, of New Delhi (28.6139° N, 77.2090° E) region, India. Several classes were identified like forest, parking, residential areas, slums, wasteland, water bodies. It was observed that our proposed framework outperformed with classification accuracy of 89.43%, compared to the current state-of-the-art algorithms (support vector machine (SVM), K-nearest neighbor (KNN), and random forest (RF)). Accuracy assessment was done by means of following statistic measures (precision, recall, specificity, and area under curve (AUC)) and receiver operating characteristic (ROC) curve.

Keywords Land use-land cover · Convolutional neural network
Sentinel 2 · Deep learning

R. Nijhawan
Indian Institutes of Technology, Roorkee, 247667 Roorkee, India
e-mail: rahul.deq2014@iitr.ac.in

D. Joshi (✉) · N. Narang · A. Mittal · A. Mittal
Graphic Era University, 248002 Dehradun, India
e-mail: deepankar23.j@gmail.com

N. Narang
e-mail: narang28naman@gmail.com

A. Mittal
e-mail: mittal.adi333@gmail.com

A. Mittal
e-mail: dr.ankush.mittal@gmail.com

© Springer Nature Singapore Pte Ltd. 2019

87

J. K. Mandal et al. (eds.), *Advanced Computing and Communication Technologies*,
Advances in Intelligent Systems and Computing 702,
https://doi.org/10.1007/978-981-13-0680-8_9

1   Introduction

The inordinate advancement in the remote sensing technology has laid down a solid foundation for the researchers all around the globe to analyze and prospect the cosmic possibilities and services we can offer to the people who need the revolution to happen and are counting on it. Prodigious amounts of data (i.e., images with various imaging modalities, spatial and spectral resolutions, and dynamic ranges) increase the challenges to automatically analyze the data and select the helpful information from them.

In recent years, the growth of image processing in remote sensing technology for low-level as well as high-level tasks, such as denoising or classification has led to availability of surfeits of land cover classification algorithms which are developed with firm theoretical substratum based on spatial and spectral pixel properties. Still moving from pixels to images and then to scenes require more acute efforts.

Dynamic identification of any pixel or scene is necessary as no data can be processed in the binary world unless it is tagged to a particular identifier. Hence, classifying an image according to a set of semantic categories is the goal of some of the classifications. The problem is complex because land cover characterization of an appropriate class may present a large variability and objects may appear at different scales and orientations. Another problem which we generally face same land cover and even same articles can be found in imagery related to disparate classes.

Low-level features typical of pixel-based or object-based approaches [1–3], become abortive encoding spectral, geometrical and textural properties. There is a viable need of more detailed features and descriptors to seize the linguistics of the scene.

Moreover, the affluent and attractiveness of descriptors proved to be a pathfinder for deep neural networks. Deep learning has brought out magnificent outcomes in object recognition [4]. It has also been tested in remote sensing tasks [5–7] including land use pixel classification, always displaying a tremendous prospective. The evidence of the superiority of deep learning can be justified by its ability to classify among very tenuous differences.

In this paper, a new deep learning framework approach of ensemble of CNNs has been proposed. Each CNN is inputted with different combinations of spectral bands, topographic, and texture parameters. An integrated feature vector is composed of the individual feature vectors obtained from different CNNs. The results of our proposed approach outperformed when compared with the current state-of-the-art algorithms. Accuracy assessment was accomplished by means of several statistic measures and ROC curves.

The organization of the paper is as follows: Sect. 2 (Related Work), Sect. 3: discusses the dataset size and resources, Sect. 4: discusses methodologies used in our work, Sect. 5: discusses the proposed framework. In Sect. 6, we will discuss the results obtained and draw conclusions (in Sect. 7) from evaluated experiments.

2   Related Work

Land use-land cover classification is a captivating aspect of how we apply deep learning in our day-to-day lifestyle where semantic classification of various study areas is performed using CNN [8]. Using pre-trained CNN is an intelligent way to obtain best results among state of the art [8], where models like GoogleNet and CaffeNet were tested [8], and intense research is being carried out on classification of remote sensing scene, targeting both on the usage of proper classification tasks and of applicable image descriptors. Among local descriptors, local binary pattern (LBP) [9], scale invariant feature transform (SIFT) [10], or histogram of oriented gradients (HOG) [11] have proved to be effective in various computer vision applications, especially in object recognition. Spatial pyramid match kernel (SPMK) [12] matches the local features of a partitioned image at different levels of resolution and further computes the weighted histograms at each level. Apart from grayscale images, Yang and Newsman [13] showed that color histogram descriptors evaluated on hue, lightness, and saturation (Color-HLS) yielded extraordinary results in comparison to efforts done at time of the then state of the art.

The current state of the art is the contribution of some very esteemed researchers which started with inspected pixel-based mechanisms for land cover classification [14] where support vector machine (SVM) resulted in accuracy of about 75%, the neural network (NN)-based classifier gave an overall accuracy (OA) of about 74% with same efficiency. But the SVM resulted in consumption of too many resources in big data applications. Hence, a popular random forest (RF) [15]-based approach was adopted; however, for an efficient use of random forest approach, multiple features need to be wangled to operate the random forest classifier.

Then the interest shifted toward ensemble-based [16–19] multisensory and multi-temporal land cover classification methods which proved to be very efficient and outperformed SVM [20–22].

Deep learning (DL) proved to be a very powerful tool for solving an enormous set of problems which we usually deal with in computer vision, signal processing, and natural language processing [23]. Deep learning is based on simulating the human vision to solve big data problems in computer vision and provides semantic information at the output. Deep learning proved to be an efficient tool for processing radar, hyperspectral, and multispectral images and for extracting land cover types such as vegetation cover, road extraction, building extraction. [24, 25]. Several deep learning architectures have been used like convolutional neural networks, deep autoencoders, deep belief networks, and recurrent neural networks for remote sensing task [25–28].

Table 1 Dataset sources

Satellite imagery	ID	Coordinates	Acquisition date
Sentinel 2	L1C_T43RGM_A010069_20170527T052847	28.4078758, 77.6014517	2017/05/27
	L1C_T43RGM_A009969_20170520T053057	28.4078758, 77.6014517	2017/05/20
	L1C_T43RGM_A009783_20170507T052840	28.4078758, 77.6014517	2017/05/20

Table 2 Classes in the dataset

Class name	Number of images
Airports	232
Building	356
Forest	298
Parking areas	276
Slums	356
Water bodies	409
Total	**1927**

3 Data Collection

Sentinel 2 satellite image dataset of spatial resolution of 10 m for the year 2017 is used in our work for land use/cover classification. The dataset was obtained by the courtesy of United States of Geological Survey (U.S.G.S) (http://earthexplorer.usgs.gov/) depicted in Table 1. The dataset was imported into ERDAS Imagine Version 10.1 (Leica Geosystems, Atlanta, USA), for satellite image preprocessing. The data was used with wavelengths ranging between 20 nm (Band-8A) and 115 nm (Band-8). We created the dataset of 1927 images showed in Table 2 where ratio of 4:1 between training and validation phase was maintained, with the following classes: forests, airports, buildings, parking areas, slums, and water bodies.

4 Methodology

4.1 Support Vector Machine (SVM)

SVM is supervised learning algorithms that analyze data for classification and regression analysis. It makes a supervised data classification model to a non-probabilistic binary linear classifier. When data are not supervised, it attempts to find natural clustering of data which is also known as support vector clustering and it has a very wide use in industrial projects [21, 22].

4.2 Convolutional Neural Network (CNN)

CNN aims at reproducing behavior of human brain by labeling images through a black box deep neural network that tries to replicate cerebral mechanism combining, internal descriptions of increasing level of abstraction. Nonetheless, deep learning is most efficient way to solve our problem today with the best accuracy. Previous work [29] proved that convolutional neural networks proved to be one of the most powerful networks among the state of the art. But the only drawback that is considered while using CNN is the requirement of large dataset.

5 Proposed Architecture

In this paper, we have proposed a new deep learning framework model that contains three individual CNNs ensembled together. The following inputs have been passed to the CNNs. CNN1: Sentinel 2 (Bands, B2, B3, B4, B8, and B8A), CNN2: Sentinel 2 (Bands B11, B12), and CNN3: topographic parameters (Shape, Aspect, Surface Curvature) and texture parameters (Variance, Entropy, Dissimilarity). Each CNN model used the AlexNet pre-trained architecture on ImageNet dataset [4], which is fine-tuned on our created dataset (Sect. 3). Then we integrated the output feature vectors from individual CNNs, forming a combined feature vector F (F1 + F2 + F3) for our ensemble framework. Further we employed SVM model for land use-land cover classification. Our proposed new deep learning framework is represented in Fig. 1

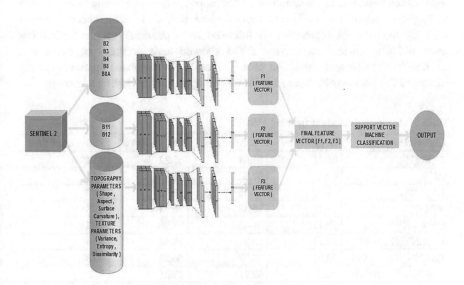

Fig. 1 Proposed DL framework (ensemble of CNNs) for land use-land cover classification

5.1   Brief Description of the AlexNet Architecture

It consists of eight weighted layers which are further grouped into first five being convolutional layers and the last three being fully connected layers.

First layer takes input $224 \times 224 \times 3$ images as input with kernel size $11 \times 11 \times 3$ (total 96), a tread of four pixel (i.e., seperation between amenable fold controls of contiguous neurons in the kernel map) and filters it to provide response normalized and pooled output as an input to the adjacent layer. Second convolutional layer takes input from the first convolutional layer and strains it with 256 number of kernels of sized $5 \times 5 \times 48$. Kernels of second, fourth and fifth convolutional layer are anchored to all kernel maps in the second layer.

Response normalization layer follows first and second convolutional layer, whereas max pooling layer follows both normalization layers as well as fifth convolutional layer. Third convolutional layer has 384 number of kernels sized $3 \times 3 \times 256$ anchored to the outputs of second convolutional layer (i.e., normalized and pooled). Fourth convolutional layer has 384 number of kernels sized $3 \times 3 \times 192$. Fifth convolutional layer has 256 kernels of size $3 \times 3 \times 192$. Third, fourth, and fifth convolutional layer are connected to one another without any intervening pooling of normalization layers. Fully connected layers have 4096 neurons each where every layer is connected to the other two layers.

6   Results and Discussion

This section presents the results of the tests conducted during the course of our work which validated our assumption of the proposed approach being most efficient among the state of the art. Tables 3 and 4 show precision, recall, specificity, and AUC for the different approaches we followed in the previous sections for training and validation phase, respectively. SVM showed least accuracy of 61.21 and 65.32% in training and validation phase, whereas our proposed approach clearly surpassed the previous best (i.e., random forest approach) by giving an accuracy of

Table 3 Accuracy assessment (training phase)

Classification	Accuracy (%)	Specificity (%)	Sensitivity (%)	Kappa coefficient
SVM [15]	61.21	64.64	67.22	0.61
ANN	64.32	67.32	66.43	0.66
KNN [16]	49.22	48.31	53.42	0.48
RF [17]	76.21	70.33	76.53	0.73
CNN + SVM [18]	75.11	71.42	74.36	0.73
S1 (CNN(1L) + RF)	77.13	75.55	78.42	0.74
S2(CNN(2L) + SVM	79.31	77.52	80.28	0.78
PA(CNN3L + SVM)	86.23	82.11	87.21	0.85

Table 4 Accuracy assessment (validation phase)

Classification	Accuracy (%)	Specificity (%)	Sensitivity (%)	Kappa coefficient
SVM [15]	65.32	68.22	71.32	0.64
ANN	65.32	68.32	68.91	0.69
KNN [16]	52.11	50.98	56.12	0.52
RF [17]	78.32	72.32	78.43	0.74
CNN + SVM [18]	78.32	73.22	76.32	0.75
S1 (CNN(1L) + RF)	81.32	75.23	79.32	0.76
S2(CNN(2L) + SVM	84.32	79.21	82.32	0.81
PA(CNN3L + SVM)	**89.43**	**84.42**	**90.22**	**0.86**

86.23% in training and 89.43% in validation with a specificity of 84.42% and sensitivity and AUC of 90.22% and 0.91, respectively. Tables 3 and 4 show the accuracy assessment of various scenarios created during the experimental phase and are mentioned as follows.

Scenario 1 (2CNNs + RF). Sentinel 2 satellite imagery spectral bands (topography and texture) parameters are provided as input to the single CNN layer, and output is classified using RF, KNN, ANN, and SVM where RF showed highest classification accuracy.

Scenario 2 (3CNNs + SVM). In this scenario, we used 2 CNN layers Sentinel 2 satellite spectral band in layer 1 and combination of textural and topographic bands in layer 2 was given as input then output from each band is combined to form a feature vector which is again tested with state-of-the-art algorithms where SVM showed maximum classification accuracy.

Scenario 3 (4CNNs + SVM). Here we employed 3 CNNs where each layer was provided with input of Sentinel 2 satellite imagery spectral bands layer 1(B2, B3, B4, B8, B8A) bands, layer 2(B11, B12) bands, and layer 3 with combination of textural and topographic bands. The output of CNNs is then combined to form a feature vector F which is further classified with state-of-the-art algorithms where SVM proved to be most efficient and provided us the highest classification accuracy.

In further experiments when we increased the number of CNN layers, the training time increased 4–5 times but the increase in accuracy was very low; hence to obtain maximum efficiency, we limited our model to 3 CNN layers.

ROC curves in Fig. 2 are plotted to estimate the performance of our proposed model. The graph is plotted between sensitivity and 100-specificity being on y- and x-axis, respectively. AUC values depicting area under the curve are used to evaluate the performance of our predictive models. The degree of perfection of model can be estimated by closeness of AUC values to 1.

The results of the ROC curves proved the proposed approach to be most efficient (with AUC = 0.91) which clearly outperformed AUC values of SVM, ANN, KNN, RF, CNN + SVM being 0.73, 0.81, 0.87, 0.86, 0.87, respectively. Interesting trend that we observed was that by increasing number of CNN layers, there was a steady

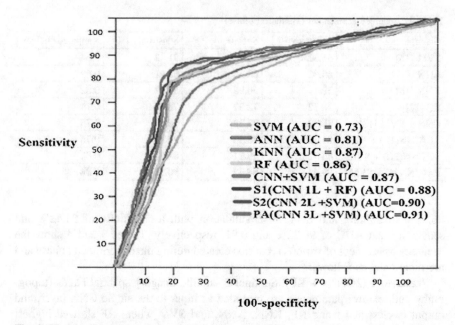

Fig. 2 Model validation by means of ROC curves and AUC analysis for the proposed predictive models where, *KNN* K-nearest neighbor, *RF* random forest, *(CNN + SVM)* convolutional neural network + support vector machine, *S1(CNN1L + RF)* single CNN with RF, *S2(CNN2L + SVM)* ensemble of two CNN's with SVM and *PA(CNN3L + SVM)* ensemble of three CNN's with SVM

increase in AUC values when we shifted from scenario 1(S1) to scenario 2(S2) and then to scenario 3(PA) but further increase in CNN layers also increased training time to a larger extent and the increase in AUC value was very less. Hence, we can come to the conclusion that our proposed approach (ensemble of 3 CNNs + SVM) proved to be most efficient model for the land use-land cover classification.

7 Conclusion

We addressed the problem of land use-land cover classification with the help of deep convolutional neural network where the use of ensemble of CNNs was limited only to feature extraction, but yielded very good result among the state of the art and surpassed the previous methods with respect to both speed and efficiency. A major challenge that was faced during the experimental phase was the shortage of data. Hence, pre-trained AlexNet model was fine-tuned which improved the efficiency of our work.

Satellite images were able to perform really well in our work which also opens doors for testing of other types of images with different pixel configurations. Better performance was seen by the use of ensemble of CNNs as expected. Hence, it is clearly evident that deep neural network can be used to solve many problems we face in field of remote sensing.

References

1. Pesaresi, M., Gerhardinger, A.: Improved textural built-up presence index for automatic recognition of human settlements in arid regions with scattered vegetation. IEEE J. Sel. Top. Appl. Earth Observations Remote Sens. **4**(1), 16–26 (2011)
2. Rizvi, I.A., Mohan, B.K.: Object-based image analysis of high-resolution satellite images using modified cloud basis function neural network and probabilistic relaxation labeling process. IEEE Trans. Geosci. Remote Sens. **49**(12), 4815–4820 (2011)
3. Gaetano, R., Masi, G., Poggi, G., Verdoliva, L., Scarpa, G.: Marker-controlled watershed-based segmentation of multiresolution remote sensing images. IEEE Trans. Geosci. Remote Sens. **53**(6), 2987–3004 (2015)
4. Krizhevsky, A., Sutskever, I., Hinton, G.E.: Imagenet classification with deep convolutional neural networks. Adv. Neural Inf. Process. Syst. (2012)
5. Midhun, M.E., Nair, S.R., Prabhakar, V.T., Kumar, S.S.: Deep model for classification of hyperspectral image using restricted Boltzmann machine. In: Proceedings of the 2014 International Conference on Interdisciplinary Advances in Applied Computing. ACM (2014)
6. Chen, Y., Zhao, X., Jia, X.: Spectral–spatial classification of hyperspectral data based on deep belief network. IEEE J. Sel. Top. Appl. Earth Observ. Remote Sens. **8**(6), 2381–2392 (2015)
7. Li, T., Zhang, J., Zhang, Y.: Classification of hyperspectral image based on deep belief networks. In: IEEE International Conference on Image Processing (ICIP), IEEE, 2014
8. Castelluccio, M., Poggi, G., Sansone, C., Verdoliva, L.: Land use classification in remote sensing images by convolutional neural networks. (2015). arXiv https://arxiv.org/pdf/1508. 00092
9. Szegedy, C., Liu, W., Jia, Y., Sermanet, P., Reed, S., Anguelov, D., Erhan, D., Vanhoucke, V., Rabinovich, A.: Going deeper with convolutions. In: Proceedings of the IEEE Conference on Computer Vision and Pattern Recognition (2015)
10. Ojala, T., Pietikainen, M., Maenpaa, T.: Multiresolution gray-scale and rotation invariant texture classification with local binary patterns. IEEE Trans. Pattern Anal. Mach. Intell. **24**(7), 971–987 (2002)
11. Lowe, D.G.: Distinctive image features from scale-invariant keypoints. Int. J. Comput. Vision **60**(2), 91–110 (2004)
12. Yang, Y., Newsam, S.: Bag-of-visual-words and spatial extensions for land-use classification. In: Proceedings of the 18th SIGSPATIAL International Conference on Advances in Geographic Information Systems. ACM (2010)
13. Dalal, N., Triggs, B.: Histograms of oriented gradients for human detection. In: IEEE Computer Society Conference on Computer Vision and Pattern Recognition, vol. 1, CVPR 2005. IEEE (2005)
14. Lazebnik, S., Schmid, C., Ponce, J.: Beyond bags of features: spatial pyramid matching for recognizing natural scene categories. In: IEEE Computer Society Conference on Computer Vision and Pattern Recognition, Vol. 2. IEEE (2006)
15. Khatami, R., Mountrakis, G., Stehman, S.V.: A meta-analysis of remote sensing research on supervised pixel-based land-cover image classification processes: general guidelines for practitioners and future research. Remote Sens. Environ. **177**, 89–100 (2016)

16. Gislason, P.O., Benediktsson, J.A., Sveinsson, J.R.: Random forests for land cover classification. Pattern Recogn. Lett. **27**(4), 294–300 (2006)
17. Chen, Y., Lin, Z., Zhao, X., Wang, G., Gu, Y.: Deep learning-based classification of hyperspectral data. IEEE J. Sel. Top. Appl. Earth Observ. Remote Sens. 7(6), 2094–2107 (2014)
18. Zhao, W., Du, S.: Learning multiscale and deep representations for classifying remotely sensed imagery. ISPRS J. Photogrammetry Remote Sens. **113**, 155–165 (2016)
19. Kussul, N.N., Lavreniuk, N.S., Shelestov, A.Y., Yailymov, B.Y., Butko, I.N.: Land cover changes analysis based on deep machine learning technique. J. Autom. Inf. Sci. **48**(5) (2016)
20. Kussul, N., Shelestov, A., Basarab, R., Skakun, S., Kussul, O., Lavrenyuk, M.: Geospatial intelligence and data fusion techniques for sustainable development problems. In: Proceedings of ICTERI, pp. 196–203 (2015)
21. Ding, J., Chen, B., Liu, H., Huang, M.: Convolutional neural network with data augmentation for SAR target recognition. IEEE Geosci. Remote Sens. Lett. **13**(3), 364–368 (2016)
22. Huang, F.J., LeCun, Y.: Large-scale learning with svm and convolutional for generic object categorization. In: IEEE Computer Society Conference on Computer Vision and Pattern Recognition, vol. 1. IEEE (2006)
23. Ishii, T., Nakamura, R., Nakada, H., Mochizuki, Y., Ishikawa, H.: Surface object recognition with CNN and SVM in Landsat 8 images. In: 14th IAPR International Conference on Machine Vision Applications (MVA). IEEE (2015)
24. LeCun, Y., Bengio, Y., Hinton, G.: Deep learning. Nature **521**(7553), 436–444 (2015)
25. Mnih, V., Hinton, G.E.: Learning to detect roads in high-resolution aerial images. In: Computer Vision–ECCV 2010, pp. 210–223 (2010)
26. Geng, J., Fan, J., Wang, H., Ma, X., Li, B., Chen, F.: High-resolution SAR image classification via deep convolutional autoencoders. IEEE Geosci. Remote Sens. Lett. **12**(11), 2351–2355 (2015)
27. Liang, H., Li, Q.: Hyperspectral imagery classification using sparse representations of convolutional neural network features. Remote Sens. **8**(2), 99 (2016)
28. Lyu, H., Lu, H., Mou, L.: Learning a transferable change rule from a recurrent neural network for land cover change detection. Remote Sens. **8**(6), 506 (2016)
29. Ouyang, W., Wang, X.: Joint deep learning for pedestrian detection. In: Proceedings of the IEEE International Conference on Computer Vision (2013)

Prediction Model for Crowdfunding Projects

Jaya Gera and Harmeet Kaur

Abstract Funds are the core elements of crowdfunding that draws on small contributions from relatively large number of contributors using the Internet. Success of crowdfunding efforts is crucial for their survivals. Assessing project outcome in early phases may help all the stakeholders in making informed decisions. Considering this, the proposed work suggests a prediction models that help in assessing projects pre-launch, at launch and post-launch. Models work with various project features to generate predictions. Pre-launch model uses features that are available prior to or at the time of launch of a campaign. Post-launch model is designed for prediction after launch that uses features generated after launch of a campaign. Combined model too works for post-launch prediction and uses the probability of success provided by the first and second models to find aggregate impact on success outcome of a campaign. The proposed technique exhibits that the combined model helps in improving the prediction accuracy during the initial phase consequently achieving an accuracy of 88% within 20% of the funding cycle.

Keywords Crowdfunding · Kick-starter · Success prediction · Machine learning
Success factors · Time series data

1 Introduction

In crowdfunding, a project creator launches a project on a crowdfunding site and asks for funds online. Each project has a well-defined goal to be achieved within a stipulated time period. If the goal is achieved within deadline, then the project is

J. Gera
Department of Computer Science, Shyama Prasad Mukherji College,
University of Delhi, New Delhi 110026, Delhi, India
e-mail: jayagera@spm.du.ac.in

H. Kaur (✉)
Department of Computer Science, Hansraj College,
University of Delhi, New Delhi 110007, Delhi, India
e-mail: hkaur@hrc.du.ac.in

© Springer Nature Singapore Pte Ltd. 2019   97
J. K. Mandal et al. (eds.), *Advanced Computing and Communication Technologies*,
Advances in Intelligent Systems and Computing 702,
https://doi.org/10.1007/978-981-13-0680-8_10

successful. In all-or-nothing model, creator then receives the amount and project is executed and funders are awarded their rewards. But if a project fails, creator receives nothing; i.e. funds are not rendered and no work is carried out [1].

Although, in case of failure, no exchange of money takes place [2–4], creator's time, money and the efforts spent in the entire exercise of planning, designing, launching and promoting a project are in vain. Similarly, backers' time and effort in identifying, sharing and promoting these projects go waste [4]. So, it is in creators' and backers' interest to know the probability of project success as early as possible. Moreover, knowing project outcome in advance helps the users in working out future strategies [2]. Creators, whose campaigns have low probability, may want to improve their chances by providing updates, improving reward levels, by promoting through social media and other means. Creators, whose campaign has higher chance to receive funds, may start their future action plan in order to deliver in time [2]. Similarly, backers may promote the campaign among friends, if probability is low and may adjust their pledge if a project is lagging by little amount to make the project successful [5].

Creator launches a campaign by creating a project web page for the campaign on the crowdfunding site. Project page is a well-defined structured page that outlines objective, category, goal amount, deadline, product description, associated risks and challenges, commitments, reward tiers, etc. The information communicated via the project page has an effective and valuable contribution on project outcome [6] and provides help to backers in taking decision about funding. These features too characterize the campaign. So, these features of projects are used by the predictor model to evaluate success probability of a project. One more dimension of measuring success is how the project is performing during funding cycle, i.e. analyzing funding pattern. Successful and unsuccessful projects do follow a different pattern. Successful projects start accumulating funds from the beginning, whereas unsuccessful projects fail to attract funds from the beginning itself. Analyzing funding pattern helps in predicting project outcome at early stages, which is crucial for project success, as creator/funders can be well prepared and take appropriate actions/decisions. The aim of proposed work is also to predict project success at an early stage to assist funders/creators, so that they can make best use of their resources.

Proposed work suggests a prediction model that evaluates projects in three stages: pre-launch, at launch and post-launch. Prediction model uses various project features for generating predictions. These features vary in nature and are available at different stages of a campaign. Pre-launch/launch prediction phase uses features that are available at the time of launch or prior to launch. They are less likely to be modified after launch, i.e. somewhat stable, and may be interpreted as static in nature. Post-launch phase uses features such as pledge amount that are generated during funding cycle, i.e. available after launch of project. They change frequently, i.e. are dynamic in nature and are time dependent. Prediction model has three submodels:

- Pre-launch predictor for pre-launch and launch time evaluation,
- Post-launch predictor for post-launch evaluation,
- Combined predictor for analyzing combined impact of above two predictors.

Pre-launch predictor provides an accuracy of 74%. This accuracy improves up to approximately 98% during funding cycle using post-launch and combined predictor. The result shows that project's features are powerful predictors; combining them with the time series characteristics like pledge status, number of backers and updates help in improving the accuracy of prediction.

This work empirically establishes that the combined prediction helps in improving initial (i.e. at the launch time) prediction accuracy from 74 to 76% and then on moving to high accuracy levels (approximately 84% within 5% time period of cycle) as few pledges arrive. The rest of the paper is structured as follows: Sect. 2 mines the literature, Sect. 3 presents dataset, Sect. 4 discusses proposed work, Sect. 5 analyses the performance of the model proposed. Conclusions are drawn in Sect. 6.

2 Literature Review

Crowdfunding is almost as old as the existence of human society, but the development of modern technology has made it more conspicuous and has fascinated people across the world with diverse interest. That's why it has drawn attention of scholars from different streams such as economics, finance, management, commerce, psychology, technology, social science, entrepreneurship.

Greenberg et al. [7] applied various machine learning techniques on features available at the initial set-up of project to predict success giving accuracy of 68%. Etter et al. [2] proposed prediction models that used time series pledge money data, twitter data and project/backer graph providing accuracy of more than 76%. Mollick [6] performed in-depth analysis of factors leading to success and role of geography in deriving success. This study also provided insights of delay in project deliveries. Chen et al. [4] too developed a prediction model using parameters such as presence of YouTube video, count of tweeter link share and other project features using SVM model. Mitra and Gilbert [3] extracted linguistic features from project description and correlated them with success and suggested that language used by the creator to pitch their project plays vital role in making a project successful and devised a prediction model based on phrases, words used in project description along with other project features. Xu et al. [8] predicted using project updates provided by creator on kick-starter project page to keep backers updated about project status and activities. Chung [9] performed user and projects analysis and suggested a prediction model based on project, user, temporal and twitter features. Dey et al. [10] investigated effect of videos on success of campaigns from various categories. Lu et al. [11] correlated promotional activities on social network with success. They demonstrated that the promotional campaigns on social media help in improving the accuracy of predicting eventual outcomes and identifying popularity of a project.

Other dimensions too have been explored by research scholars. Belleflamme et al. [12] compared pre-order and equity crowdfunding and established using a mathematical model that increase in capital requirement results in distortion in price discrimination. They extended study to learn the impact of quality on uncertainty and information asymmetry. Koch and Cheng [13] performed quantitative and qualitative analysis and found that qualitative factors have a rational and significant influence on crowdfunding success.

Proposed work explores various campaign features and evaluates success probability pre-launch and post-launch. Study proposes a model consisting of family of three predictors that helps in improving prediction accuracy. The work also shows that the analysis of performance history of a project is not necessary to predict its outcome.

3   Dataset Collection and Pre-processing

Experiments were performed on kick-starter projects' sample dataset provided by Kickspy website. This dataset consists of data about projects and funding history of projects. Project data contains almost all the static characteristics such as category, subcategory, goal, duration of the project available on the project page. Creators from different countries such as US, UK, Australia launch projects on the kick-starter platform. So, the currency of the goal amount may be different for creators from different countries. For processing, goal amounts in currencies other than USD were converted to USD using rate of exchange of that period to make them comparable. Creator's information such as number of projects launched with the platform earlier and number of project backed was also collected from the kick-starter website for the analysis purpose.

Funding history includes each day's pledge status, i.e. total amount raised till that day for each project in the sample dataset. This data also consists of total number of backers who pledged that amount on the respective day. Total number of pledges and backer status sampled depend on the duration of a project and vary for each project, so, in order to perform experiment, status was resampled for each project to obtain equal number of samples for all projects. As maximum duration of a project in this dataset is 60 days, total number of samples was fixed at 60. The time of each status is normalized with respect to the campaign's launch date and campaign duration and defined as relative time (percentage of funding time passed) since the day of launch. Also, amount pledged was normalized to have uniformity of the status. To obtain normalized amount, the status of money pledged at each state was divided by the goal amount, i.e. normalized amount actually tells what percentage of goal amount is raised till that day.

In addition to above data, prediction model involves two more dynamic features: status of number of updates and comments on each day of the funding period.

To add this information to the sample dataset, updates and comments page for each project were retrieved by running a crawler, then comments and updates of each project were extracted from the crawled pages and were assembled in csv files. For each project, comment file stores name of commenter, date and time of comment and comment text; and update file stores update title, date and time of update and update text. These csv files were further processed to have count of updates and comments uploaded each day of project cycle. These files were further processed and transformed to a matrix format that have equal number of status (total 60 samples) for each project.

This sample data contains 4,682 projects launched in the month of April 2014. This dataset has successful, live, cancelled, suspended and unsuccessful projects. As study focuses on understanding characteristics of successful and unsuccessful projects, dataset was filtered to have successful and unsuccessful projects only. Out of 4,682 projects, there are total 4,121 such projects. Out of 4,121 projects, 1,899 (46%) are successful and 2,232 (54%) are unsuccessful. This dataset comprises of projects in all fifteen categories as classified by kick-starter such as music, dance. In this dataset, 823 (approximately. 20%) projects are from film and video category; 610 (14.8%) from music, 454 (11%) projects from publishing.

4 Proposed Model

The sample dataset is rich and balanced in project features required for performing analysis. All the features are not available prior to launch or at the time of launch. So based on the availability of features and time when predictor should be used to assess success, a prediction model is developed. The proposed model uses the following set of predictors to evaluate a project's probability of success:

- Pre-launch predictor,
- Post-launch predictor,
- Combined predictor.

Pre-launch predictor uses project features such as category, subcategory, goal to predict success probability. Post-launch predictor uses status available at time t such as pledge status, number of backers. Combined predictor combines the above two predictors. Logistic regression is applied to all the three predictors to evaluate success probability.

4.1 Pre-launch/Launch Predictor

Pre-launch predictor's decision is based on project features that are independent of the funding cycle and are available at the time of project launch. On kick-starter, each project is characterized by a number of features. Each project has a target amount and an end date to achieve it. A project is launched under any one of fifteen categories as defined by kick-starter. Each category has been further categorized to subcategories; for example, film and video have been classified as action, animation, documentary, etc. A creator also shares project planning, team members, expected delivery, possible delays, work experience, etc., via project page. Videos and images on project page promote and enhance project presentation. Page also depicts creator's profile such as number of Facebook friends, projects created and backed earlier. This dataset does not contain image, videos, and project description uploaded on project page, but does provide useful statistics such as number of images, videos, words in description and other statistics of projects. All these features of projects used by pre-launch predictor to evaluate success probability of a project p are summarized in Table 1.

Pre-launch probability PRE(p) for a project p is obtained by estimating the log of odds ratio using logistic regression and controlling for log of above variables. So, PRE(p) is defined as

$$PRE(p) = \frac{e^x}{1 + e^x} \qquad (1)$$

where $x = \beta_0 + \beta_{1k} * Cat(p) + \beta_{2j} * Subcat(p) + \beta_3 * Goal(p) + \beta_4 * Duration(p) + \beta_5 * Rewards(p) + \beta_6 * Conn(p) + \beta_7 * Frnds(p) + \beta_8 * HasVideo(p) + \beta_9 * Created(p) + \beta_{10} * Backed(p) + \beta_{11} * \#Video(p) + \beta_{12} * \#Image(p) + \beta_{13} * \#wordDesc(p) + \beta_{14} * \#wordRisk(p)$.

Table 1 Features/variables used by predictor

Variables	Description
Cat(p)	Category of project p
Subcat(p)	Subcategory of project p
Goal(p)	Target amount to be raised
Duration(p)	Number of days for which project is live on platform
Rewards(p)	Number of reward levels
Conn(p)	Yes, if creator shares Facebook connection
Frnds(p)	Number of Facebook friends of the creator
HasVideo(p)	Does project has any video(s)?
Created(p)	Number of projects creator has created previous to the launch of this project
Backed(p)	Number of projects creator has backed previous to the launch of this project
#Video(p)	Number of videos
#Image(p)	Number of images
#wordDesc (p)	Number of words in the description of the project
#wordRisk(p)	Number of words in the risk and challenges section

And β_0, β_{1k}, β_{2j}, ..., β_{14} are the coefficients of the independent variables. These are best-fit parameters to the logistic curve. β_{1k} represents one of the k coefficients as there are k categories ($1 \leq k \leq 15$) and β_{2j} represents one of the j coefficients for j subcategories depending upon the category and subcategory of the project.

4.2 Post-launch Predictor

Kick-starter platform provides the pledge and backer status to keep creator and backer updated about the percentage of funds raised and number of backers who have pledged that amount. Not only this, kick-starter employs different mechanisms, so that a creator can communicate with backers and keeps them informed about progress of the project. Creator can post updates to keep backers engaged and encourage them to publicize the project. Updates can be public or for only backers. Updates specific for backers are directly emailed to them. We crawled updates available publicly for all the projects in the dataset and calculated the number of updates posted each day of cycle and then resampled them to have uniform number of samples for each project. Similarly, backers can communicate with creators by posting issues, queries, responses, etc., on project page. Comments were also crawled and processed in similar way.

This information is generated during funding cycle, after a campaign is launched on the crowdfunding site. At time of launch, these are null fields and updated as pledge arrives and updates and comments are posted on the site. The campaign page provides updated values for these features at any time t. So, these time series features are used for evaluating project outcome during funding cycle, i.e. post-launch.

Post-launch predictor makes use of these dynamic, i.e. time series features and computes probability at any time t by estimating the log of odds ratio using logistic regression. So, post-launch probability $POST_t(p)$ of a project p at any time t is defined as

$$POST_t(p) = \frac{e^y}{1 + e^y} \qquad (2)$$

In (2), $y = \alpha_0 + \alpha_1 * plg_t(p) + \alpha_2 * bck_t(p) + \alpha_3 * upd_t(p) + \alpha_4 * comm_t(p)$.
And α_0, α_1, ..., α_4 are the coefficients of the independent variables.

Here, $plg_t(p)$ defines normalized value of pledge status, $bck_t(p)$ represents number of backers pledged, $upd_t(p)$ tells total number of updates uploaded by creator and $comm_t(p)$ gives total number of issues, responses, queries raised by backers at time t for a project p.

4.3 Combined Predictor

To have a better picture of the project performance and to understand combined impact of all these features, the two predictions are combined. Combined predictor combines the probability obtained from above two predictors. Logistic regression is applied on probability generated by pre-launch predictor and post-launch predictor at any time t, and a new probability CPR_t is obtained for a project p.

$$CPR_t(p) = \frac{e^z}{1 + e^z} \tag{3}$$

In (3), $z = \gamma_0 + \gamma_1 * PRE(p) + \gamma_2 * POST_t(p)$ and $\gamma_0, \gamma_1, \gamma_2$ are the coefficients of the independent variables.

A combined model may be designed to assess campaign by putting all the features available at time t on the project page in one go, but this part of proposed model uses probability of first two submodels to predict because pre-launch predictor submodel prediction output is available as soon as project is launched and as features are more of static nature; it is of little use to re-compute it. Post-launch predictor can work fast with the less features. Also, first two models can work independently. Combined predictor is designed to utilize the output available from the two models.

5 Performance Analysis

A number of techniques were applied and compared to search for an efficient method to perform both pre- and post-launch prediction. Best results were obtained using logistic regression, so it was chosen for conducting the experiments. The performance of three predictors is shown in Fig. 1.

5.1 Pre-launch Predictor Results

To train pre-launch predictor, a number of methods were employed on dataset using SPSS. Models were created using logistic regression, classification tree with methods CHAID, and CRT, and discriminant analysis. Out of all these models, the performance of logistic regression had the highest accuracy.

The accuracy of pre-launch prediction using logistic regression is 74% (shown in Fig. 1). Pre-launch predictor evaluates failed cases more accurately with the accuracy of 77.1%; i.e. model is helpful in identifying projects that need to be improved. Pre-launch predictor provides better accuracy than model proposed (accuracy was 68%) by Greenberg et al. [7].

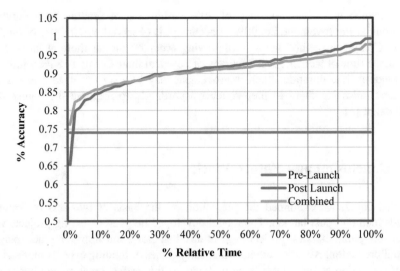

Fig. 1 Accuracy of three predictors

5.2 Post-launch Predictor Results

To find the best model, models were developed using linear discriminant analysis, *K*-nearest neighbours, classification and regression trees, Gaussian Naive Bayes, support vector machines and logistic regression. Best performance was provided by logistic regression. So, post-launch predictor was trained and tested by applying logistic regression using 10 fold cross validation in Python. Initial accuracy reaches 85% accuracy within 10% of funding time. It performs better than the model proposed by Etter et al. [2] that achieves 85% accuracy in 15% of funding time. Prediction accuracy drastically improves as time passes reaching 90% within 30% of funding time period (shown in Fig. 1).

5.3 Combined Predictor

Pre-launch predictor uses the campaign features and provides accuracy of 74%. This initial accuracy can still be improved. The combined model helps in improving this initial probability, i.e. probability just after launch. The first few pledges play a significant role in enhancing prediction accuracy, which moves from 76 to 84% (shown in Fig. 1). Combined model has improved accuracy, precision and recall over dynamic model.

Pre-launch predictor maintains its accuracy of 74% during the entire cycle as features remain same, so is probability. Post-launch predictor initially gives an accuracy of 65% and grows drastically with the knowledge of few first pledges just after launch and gives accuracy of 80% within 3% of funding time duration. Substantial improvement, i.e. 90% of accuracy, is observed within 30% of funding time. Combined method helps in improving accuracy further more at initial time, i.e. at the time of launch, gives accuracy of 76% and increases up to 84% within 5% of funding time duration. Such improvement may be very useful to creators and backers, allowing them to prepare/react accordingly to improve the course of a campaign [2].

6 Conclusion and Future Work

In this paper, we examined the prediction on the basis of two sets of features: pre-launch and post-launch. The model developed helps in assessing projects with available parameters on project page at any instance t and improving accuracy of initial prediction. Accurate prediction at early stages of funding cycle is more useful than prediction at some later stage of cycle, as this assists creators and backers in strategizing their future action plan. Combining probability of two models helps in improving initial prediction.

This work considers features that are mainly numerical in nature. We would like to extend this work by analyzing textual data such as project description, updates, comments. and visual data such as images and videos. This work has been limited to statistics only. In future, a model with multiple analytics which also considers textual and visual content of project, social media and communication content could be developed.

Acknowledgements A portion of the dataset used in this work has been downloaded from website http://www.kickspy.com/. This was released by owner of this website. We are thankful to the owner of the website for releasing it as an open-source database.

References

1. Cumming, D.J., Gael, L., Schwienbacher, A.: Crowdfunding models: Keep-It-All vs. All-Or-Nothing. In: Finance Meeting EUROFIDAI-AFFI Paper, pp. 1–41 (2014). https://doi.org/10.2139/ssrn.2447567
2. Etter, V., Grossglauser, M., Thiran, P.: Launch hard or go home!: predicting the success of Kickstarter campaigns. In: Proceedings of the First ACM Conference on Online Social Networks—COSN'13, pp. 177–82 (2013)
3. Mitra, T., Gilbert, E.: The language that gets people to give. In: Proceedings of the 17th ACM Conference on Computer Supported Cooperative Work & Social Computing—CSCW'14, pp. 49–61 (2014)
4. Chen, K., Jones, B., Kim, I., Schlamp, B.: KickPredict: predicting Kickstarter success. Technical report, California Institute of Technology (2013)
5. Wash, R.: The value of completing crowdfunding projects. In: Proceedings of the 7th International Conference on Weblogs and Social Media, ICWSM 2013, pp. 631–39 (2013)

6. Mollick, E.: The dynamics of crowdfunding: an exploratory study. J. Bus. Ventur. **29**(1), 1–16 (2014)
7. Greenberg, M.D., Hariharan, K., Gerber, E., Pardo, B.: Crowdfunding support tools: predicting success & failure. In: CHI 2013, Changing Perspectives, 1815–20, Paris, France (2013)
8. Xu, A., Yang, X., Rao, H., Fu, W., Huang, S., Bailey, B.P.: Show me the money! an analysis of project updates during crowdfunding campaigns. In: Proceedings of the 32nd Annual ACM Conference on Human Factors in Computing Systems, pp. 591–600 (2014)
9. Chung, J.: Long-term study of crowdfunding platform: predicting project success and fundraising amount. In: 26th ACM Conference on Hypertext and Social Media, pp. 1–10 (2015)
10. Dey, S., Duff, B., Karahalios, K., Fu, W.T.: The art and science of persuasion: not all crowdfunding campaign videos are the same. In: Proceedings of the 2017 ACM Conference on Computer Supported Cooperative Work and Social Computing, pp. 755–769. ACM (2017)
11. Lu, C., Xie, S., Kong, X., Yu, P.S.: Inferring the impacts of social media on crowdfunding. In: Proceedings of the 7th ACM International Conference on Web Search and Data Mining, pp. 573–82 (2014)
12. Belleflamme, P., Thomas, L., Schwienbacher, A.: Crowdfunding: tapping the right crowd. J. Bus. Ventur. **29**(5), 585–609 (2014)
13. Koch, J., Cheng, Q.: The role of qualitative success factors in the analysis of crowdfunding success: evidence from Kickstarter. In: Proceedings of the 20th Pacific Asia Conference on Information Systems (PACIS 2016), Chiayi, Taiwan (2016)

An Enhanced Approach for Detecting Helmet on Motorcyclists Using Image Processing and Machine Learning Techniques

Abhijeet S. Talaulikar, Sanjay Sanathanan and Chirag N. Modi

Abstract In this paper, we propose an approach to automatically identify the bike riders who are not wearing helmets. This approach takes video feed from the surveillance camera deployed at roads and applies a background subtraction technique to identify moving vehicles. From the foreground blobs, different features are extracted to identify motorcycles among the other vehicles. From the motorcycle objects, the head region of the blob is considered to extract helmet-related features. For the performance and accuracy improvement, we apply principal component analysis (PCA) on the derived features. To detect helmet from the motorcycle object, we apply different machine learning techniques on the selected features and perform the feasibility analysis.

Keywords Helmet detection · Image processing · Principal component analysis
Machine learning

1 Introduction

Many of the urban cities are facing the problems of traffic congestion, violation of the traffic laws, and the increasing number of road accidents. Among numerous accidents, the motorcycle riders are at the greatest risk of injury and death due to not following safety measures like riding motorcycle without helmet. Through wearing good quality helmets, the risk of death and the risk of severe head injury can be reduced. Various traffic laws have been introduced to ensure that motorcycle riders will wear the helmets. However, these laws are not being followed by the riders, which have compelled researchers to investigate smarter methods. One solution is to design helmet with inbuilt sensors [1]. Such sensors can be used in multiple

A. S. Talaulikar (✉) · S. Sanathanan · C. N. Modi
National Institute of Technology Goa, Farmagudi 403401, India
e-mail: abhijeetstalaulikar@gmail.com

C. N. Modi
e-mail: cnmodi@nitgoa.ac.in

© Springer Nature Singapore Pte Ltd. 2019
J. K. Mandal et al. (eds.), *Advanced Computing and Communication Technologies*,
Advances in Intelligent Systems and Computing 702,
https://doi.org/10.1007/978-981-13-0680-8_11

ways; one such way is to connect the helmet with the motorcycle's engine in such a way that the engine will start only when the helmet is worn and buckled. However, this will be very expensive to manufacture such sophisticated helmets. In addition, it makes complex design of motorcycle. Another solution is the manual surveillance of vehicles at busy roads. For instance, CCTV cameras on roads can be used to monitor the traffic, identify the causes and demographics of road accidents, and take the necessary action to prevent them. However, the control crew has to manually monitor the video footages. Such an offline system requires a lot of manpower and storage for video footages.

To address above problems, we propose an efficient approach for automatically detecting helmet on motorcyclist using image processing and machine learning techniques. It takes video streams from surveillance camera deployed at roads and converts it into frames. To reduce overall computational cost, it selects only key frames with higher information to process further. From these frames, features of the foreground image are extracted to identify motorcycle objects among the other vehicle objects. The head region of the motorcycle objects is considered for the extraction of the further features related to helmet. We apply principal component analysis (PCA) for the reduction of features and to improve accuracy in helmet detection. We apply different machine learning techniques to predict the presence of helmet.

Rest of this paper is organized as follows. Section 2 discusses the existing works in the area of identifying helmets from the vehicle images, while the proposed approach is discussed in Sect. 3. Section 4 presents the experimental results of the proposed approach. Section 5 concludes our work with references at the end.

2 Related Work

2.1 Background Subtraction

Lo et al. [2] have used the median of n frames in video instead of averaging the frames for background subtraction. However, it does not consider the effect on background due to light changes. To address the problem of light changes in background subtraction, Ridder et al. [3] have applied Kalman filter to each pixel, which uses least mean square method. It considers two parameters, viz observed pixel value and predicted value. It helps in predicting the moving objects in the next frame. However, it cannot work well with continuously moving foreground objects. Wren et al. [4] have used multi-class statistical model named as Pfinder for tracking human in the images. They have modeled each pixel using a single Gaussian and derived feature vector at each pixel by adding spatial coordinates to the textual components of image. These features are clustered as per color and spatial simi-larity. Such combination forms blobs with similar image properties. Such system cannot handle frequent changes in the images. Friedman and Russell [5] have used an expectation maximization (EM) framework for detecting moving vehicles. They have classified the image pixels into three distributions, viz road color, shadow color, and vehicle color. However, some of the pixels cannot be determined by all three components.

2.2 Motorcycle Detection

Chen et al. [6] have proposed a system for classifying vehicles using Kalman filter and SVM. Here, an improved Gaussian mixture model is used to deal with changing illumination. A multi-dimensional Gaussian kernel density transform is used in the background learning procedure. Leelasantitham et al. [7] have presented a method for segmenting and classifying moving vehicles. It detects moving vehicles using a blob tracking method and extracts vehicle features such as position, length, and width. These features are applied to decision tree classifier for classifying the vehicles. Chiu et al. [8, 9] have used canny edge detector to find the edges of the likely helmet region. However, this approach relies heavily on the assumption that riders often wear helmets. Silva et al. [10] have used adaptive mixture of Gaussians (AMG) to learn the background and dynamic object segmentation for detecting moving vehicles. It extracts SURF, histogram of oriented gradients, HAAR, and linear binary patterns (LBP) descriptors from these foreground blobs and applies to multi-layer perceptron, SVM, and radial basis function networks for classifying the vehicles. Mukhtar et al. [11] have used circular Hough transform to detect circular objects such as head, helmet, or headlights. Dark vertical strips are identified as motorcycle tires. The corner points are used to find the bounding rectangles of the motorcycle blobs. Using these cues, the location of vehicle is determined and SVM classifier is trained on HOG features to verify the motorcycle. Waranusast et al. [12] have extracted features like area of the bounding rectangle, aspect ratio of the bounding rectangle, and standard deviation of hue around the blob center to identify motorcycle objects. Karaimer et al. [13] have used shape-based and gradient-based descriptors. Shape-based features such as convexity, elongation, and rectangularity are extracted to train KNN, while the gradient features like HOG are used with a SVM to identify vehicle objects.

2.3 Helmet Detection

Marayatr et al. [14] have considered the circular shape of the helmet. They used multiple threshold methods to subtract the background. Here, a multi-layer neural network is trained on the area of the rectangular blob to identify the motorcycles. To detect helmet, circular Hough transform is applied on the top part of the blob. Liu et al. [15] have detected a full-face helmet by considering circles through canny edges. This approach needs to consider other parameters since it will not distinguish the shape of a bare head and helmet. Chiverton [16] has used the reflective property of helmet. They applied an improved adaptive background model with shadow detection to subtract the background. They used SVM classifier to detect helmet. Silva et al. [17] have detected moving vehicles using the AMG algorithm. They used linear binary patterns (LBP) features to train SVM classifier in order to separate motorcycles and non-motorcycles. From the motorcycle blobs, features like circular-shaped descriptor, gradients, and a hybrid descriptor of texture and gradients are extracted and applied to a probabilistic (Naive Bayes), a geometric (SVM), and random forest classifiers to detect helmet. Doungmala et al. [18] have proposed a two-stage method to detect a helmet. In the first stage, it looks for facial features using Haar descriptor. In second stage, a circular Hough transform is used to find the helmet. Dahiya et al. [19] have used histograms (HOG), scale invariant feature transform (SIFT), and linear binary patterns (LBP) to identify both motorcycles and helmets. Waranusast et al. [20] have applied background subtraction using an improved AMG model. They have applied features such as the area of bounding rectangle, aspect ratio, and the standard deviation of hue around the center of the blob to KNN classifier for detecting motorcycle. From the head region (ROI) of motorcycles, the arc circularities, average intensities, and average hues are applied to KNN classifier to detect helmet.

The existing research efforts should not depend solely on one characteristic, e.g., circular shape or luminosity. (2) The features like circular Hough transform are computationally expensive. (3) Fast detection techniques need to be investigated.

3 Proposed Approach

The objective of the proposed approach is to detect the helmets on motorcyclist with high accuracy, minimal error rate, and low computational cost. A generic framework of the proposed approach is given in Fig. 1. It works in four phases, viz key frames selection, background subtraction, motorcycle detection, and helmet detection.

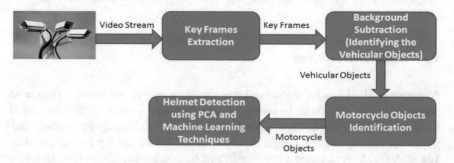

Fig. 1 Generic framework of the proposed approach

It captures the video streams form the surveillance camera which takes the front views of motorcyclist. It applies temporal differencing to these streams for selecting key frames with a significant amount of information. On the selected frames, the background subtraction technique is applied to extract vehicular objects. From these, vehicular objects, motorcycle objects are identified and finally helmet detection techniques are applied on the identified motorcycle objects.

3.1　Key Frames Selection Using Temporal Differencing

Key frames can give more information about the objects. For this, we use temporal differencing [21] which calculates the pixel-by-pixel difference between two consecutive frames. The frames satisfying the given threshold are considered as key frames. It can be calculated as:

$$D(K_i, K_{i+1}) = \sum_{x=1}^{N} \sum_{y=1}^{M} (K_i[x,y] - K_{i+1}[x,y])^2 \tag{1}$$

where D is the temporal difference, and K_i is the ith key frame. For some T, K_{i+1} is selected as key frame, if $D(K_i; K_{i+1}) \geq T$.

3.2　Background Subtraction

It is a process of extracting moving objects (foreground) from the image frames through image differencing between each frame and the reference frame (or background model). In the proposed approach, we consider an adaptive background. Here, stationary and slowly moving vehicles are not considered for the foreground since these objects lead to omission. Due to the different weather conditions (rain, snow, etc.), illumination of scenes in frames can be changed gradually. To adapt to such changes, we use adaptive mixture of Gaussians (AMG) [22] model as a background model for extracting vehicular objects. It considers the values of a particular pixel over time as a "pixel process." At any time t, particular pixel $\{x_o, y_o\}$ history is known as:

$$\{X_1, \ldots, X_t\} = \{I(x_o, y_o, i) : 1 \leq i \leq t\} \tag{2}$$

Such history of each pixel is modeled as a mixture of K Gaussians. The probability of observing the current pixel value is

$$P(X_t) = \sum_{i=1}^{k} \omega_{i,t} \times \eta\left(X_t, \mu_{i,t}, \Sigma_{i,t}\right) \tag{3}$$

where K is the number of distributions, $\omega_{i,t}$ is an estimated weight of the ith Gaussian at time t, $\mu_{i,t}$ is the mean value of the ith Gaussian at time t, and η is a Gaussian probability density function, calculated as below:

$$\eta\left(X_t, \mu_{i,t}, \Sigma_{i,t}\right) = \frac{1}{(2\pi)^{\frac{n}{2}}\Sigma^{\frac{1}{2}}} e^{\left(\frac{-1}{2}(X_t-\mu_t)^T \Sigma^{-1}(X_t-\mu_t)\right)} \tag{4}$$

A new pixel is generally represented by major components of the mixture model. Every new pixel X_t is compared against the existing k-Gaussian distributions.

3.3 Motorcycle Detection

From the background subtraction, a set of vehicle objects as foreground is extracted. To differentiate motorcycle objects from the other vehicular objects, we extract features like aspect ratio and standard deviation of hue [12] from all the vehicular objects. Aspect ratio gives the ratio between the width and the length of the minimum bounded rectangle that bounds the vehicle object. Standard deviation of hue is used to find a variation in hue around the center of vehicular object. We assume that motorcycle has more variation than other vehicles.

3.4 Helmet Detection

For helmet detection, the top 25% of the motorcycle blobs is considered as the region of interest (ROI). The ROI image is converted into binary form [23]. To reduce noise in ROI image, we apply median filtering, flood-fill, erode, and dilate as image processing (refer Fig. 2). Median filtering is used to reduce random noise by replacing each pixel with the median of the pixels in its neighborhood. Flood-fill is used to remove the isolated black spots. It raises the intensity values of dark areas to the values of the lighter areas surrounding it. Erode replaces each pixel value by the minimum value in its neighborhood. Dilate replaces each pixel value by the maximum value in its neighborhood. Erode operation followed by dilate removes small objects from the foreground, thereby considerably reducing noise. The pre-processed image is then divided into four quadrants (refer Fig. 3), and following features are extracted.

Arc circularities. We consider the left half and right half of the head region for the arc circularity measure [24]. The arc circularity C provides a measure of the arc's resemblance to the circle as given below:

Original image of head

After conversion to binary
(luminance threshold = 0.98)

After preprocessing

Fig. 2 Image preprocessing

Fig. 3 Demarcation of ROI

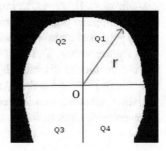

$$C = \mu_r / \sigma_r \tag{5}$$

where μ_r is the mean and σ_r is the standard deviation of the distance r from the centroid O to the boundary of the head region. We assume that a head with helmet is more circular than a head without helmet.

Average intensities. These features represent the average pixel intensities in each quadrant. It is calculated as:

$$\mu_I = \frac{1}{N} \sum_{i=0}^{N-1} I_i \tag{6}$$

where I_i is the intensity of the ith pixel in the quadrant and N is the total number of pixels in the quadrant. It is assumed that a quadrant has different average intensity range in the case of a helmet. It also represents the luminosity of the helmet.

Average hues. These features represent the average hue of the third and the fourth quadrant, as given below.

$$\mu_H = \frac{1}{N} \sum_{i=0}^{N-1} H_i \tag{7}$$

where H_i is the hue of the ith pixel in the quadrant and N is the total number of pixels in the quadrant. A head wearing a helmet reveals less skin in the third and fourth quadrants as compared to a bare head.

Histogram of Oriented Gradients (HOG). The last 1980 features are the HOG features of the top half of the head region. In our approach, we consider the first two quadrants of the 100×100 image which is reduced to size 50×100 and divided into 16×16 blocks with 50% overlap. A block consists of 2×2 cells, each with size 8×8. Thus, for a 50×100 image, we obtain a total of 55 blocks. Each histogram represents a cell size 8×8. Therefore, a block contains four histograms, each has nine components. Each block with four histograms gives a total of 36 components (4×9). Therefore, in our approach, we obtain a total of 1980 features for a given image.

Above features describe the meaningful characteristics required for further classification. However, to improve the accuracy and to reduce number of features, we have applied principal component analysis (PCA) on the derived HOG features. In the proposed approach, each image of a rider's head in the dataset is represented by a feature vector. The selected features are applied to different classifiers to map feature vectors to discrete classes, viz helmet present (positive) or helmet absent (negative). For the feasibility analysis, we test different well-known classifiers, viz logistic regression, multi-layer perceptron (MLP), support vector machine (SVM), decision tree, random forest, and k-nearest neighbor (KNN).

4 Experimental Results and Analysis

For the experimental results, we have prepared a vehicular image dataset from the motorcycle rallies and kart races obtained from the Internet. It consists of 300 images of heads wearing a helmet and 300 bare heads. For each image, we have extracted the features from motorcycle blobs, as discussed in Sect. 3.4. On the extracted features, we have applied the well-known machine learning techniques, viz logistic regression, multi-layer perceptron (MLP), support vector machine (SVM), Naïve Bayes, decision tree, random forest, and k-nearest neighbor (KNN), to test their feasibility in detecting the helmets. The experimental results are shown in Table 1. It shows that logistic regression, multi-layer perceptron, and SVM work well in detecting helmets. Based on these results, we have applied ensemble techniques, viz majority voting and weighted averaging on the output of these classifiers. The weighted averaging on logistic regression, multi-layer perceptron, and SVM classifiers performs well in identifying helmets from the road traffic images. It gives 96% accuracy with an affordable computational cost.

As shown in Fig. 4, we have plotted receiver operating characteristics (ROC) curves for weighted averaging on logistic regression, multi-layer perceptron, SVM with ten-cross fold validation. It gives larger AUC (0.99). In general, weighted averaging on logistic regression, multi-layer perceptron, and SVM classifiers performs well in terms of high accuracy (96%) and low errors (2%) in detecting the helmet on motorcyclists.

Table 1 Results of the different classifiers in identifying helmets from the images (with features reduction using PCA)

Classifier	Accuracy	Precision	Recall	F score	AUC	FN	Time (s)
Logistic regression (LR)	**0.95**	**0.95**	**0.95**	**0.95**	**0.99**	**2.13**	**0.47**
MLP	**0.94**	**0.94**	**0.94**	**0.94**	**0.98**	**2.66**	**2.19**
SVM	**0.95**	**0.95**	**0.95**	**0.95**	**0.99**	**1.83**	**1.39**
Naïve Bayes	0.90	0.90	0.90	0.90	0.96	5.89	0.23
Decision tree	0.87	0.87	0.86	0.87	0.87	6.50	0.31
Random forests	0.92	0.92	0.92	0.92	0.97	4.66	0.97
KNN	0.85	0.86	0.86	0.86	0.92	4.16	0.36
Majority voting with LR, MLP, SVM	0.95	0.95	0.95	0.95	0.99	2.66	2.41
Weighted average with LR, MLP, SVM	**0.96**	**0.96**	**0.96**	**0.96**	**0.99**	**2.00**	**4.64**

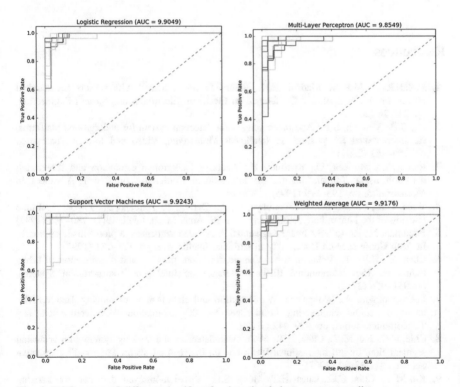

Fig. 4 ROC curves of different classifiers for detecting helmet on motorcyclist

5 Conclusions

To detect helmet on motorcyclists, we have proposed an approach which captures video streams from the front view of vehicles on the road and applies image processing and machine learning techniques. With the help of temporal differencing, we have reduced the number of frames to be processed and thus reducing the overall computational effort. For extracting the number of vehicular objects, we have applied background subtraction using an adaptive mixture of Gaussians method and differentiated motorcycle objects from the other objects by considering the appropriate parameters. For detecting the helmet from the motorcycle objects, we have extracted different features on which principal component analysis is applied, and thus, the accuracy is improved. We have analyzed different classifiers for detecting helmet and found that the experimental results of weighted averaging on logistic regression, multi-layer perceptron, and SVM classifiers performs are very encouraging.

References

1. Mohd Rasli, M.K.A., Madzhi, N.K., Johari, J.: Smart helmet with sensors for accident prevention. In: International Conference on Electrical, Electronics and System Engineering, pp. 21– 26 (2013)
2. Lo, B.P., Velastin, S.A.: Automatic congestion detection system for underground platforms. In: International Symposium on Intelligent Multimedia, Video and Speech Processing, pp. 158–161 (2001)
3. Ridder, C., Munkelt, O., Kirchner, H.: Adaptive background estimation and foreground detection using Kalman-filtering. In: International Conference on Recent Advances in Mechatronics, pp. 193–199 (1995)
4. Wren, C.R., Christopher, R., Azarbayejani, A., Darrell, T., Pentland, A.: Pfinder: real-time tracking of the human body. IEEE Trans. Pattern Anal. Mach. Intell. 19(7), 780–785 (1997)
5. Friedman, N., Russell, S.: Image segmentation in video sequences: a probabilistic approach. In: 13th Conference on Uncertainty in Artificial Intelligence, pp. 175–181 (1997)
6. Chen, Z., Ellis, T., Velastin, S.A.: Vehicle detection, tracking and classification in urban traffic. In: 15th International IEEE Conference on Intelligent Transportation Systems, pp. 951–956 (2012)
7. Leelasantitham, A., Wongseree, W.: Detection and classification of moving Thai vehicles based on traffic engineering knowledge. In: 8th International Conference on ITS Telecommunications, pp. 439–442 (2008)
8. Chiu, C.C., Ku, M.Y., Chen, H.T.: Motorcycle detection and tracking system with occlusion segmentation. In: 8th International Workshop on Image Analysis for Multimedia Interactive Services, pp. 32–32 (2007)
9. Ku, M.Y., Chin, C.C., Chen, H.T., Hong, S.H.: Visual motorcycle detection and tracking algorithms. World Sci. Eng. Acad. Soc. Trans. Electron. 5(4), 121–131 (2008)
10. Silva, R., Aires, K., Veras, R., Santos, T., Lima, K., Soares, A.: Automatic motorcycle detection on public roads. Latin-american Center Inf. Stud. (CLEI) Electronic J. 16(3), 4–4 (2013)
11. Mukhtar, A., Tang, T.B.: Vision based motorcycle detection using HOG features. In: International Conference on Signal and Image Processing Applications, pp. 452–456 (2015)

12. Waranusast, R., Timtong, V., Bundon, N., Tangnoi, C.: A computer vision approach for detection and counting of motorcycle riders in university campus. In: International Electrical Engineering Congress, pp. 1–4 (2014)
13. Karaimer, H.C., Cinaroglu, I., Bastanlar, Y.: Combining shape based and gradient-based classifiers for vehicle classification. In: 18th International Conference on Intelligent Transportation Systems, pp. 800–805 (2015)
14. Marayatr, T., Kumhom, P.: Motorcyclist's helmet wearing detection using image processing. Adv. Mater. Res. **931–932**, 588–592 (2014)
15. Liu, C.C., Liao, J.S, Chen, W.Y., Chen, J.H.: The full motorcycle helmet detection scheme using canny detection. In: 18th IPPR Conference on Computer Vision, Graphics and Image Processing, pp. 1104–1110 (2005)
16. Chiverton, J.: Helmet presence classification with motorcycle detection and tracking. IET Intel. Transport Syst. **6**(3), 259–269 (2012)
17. Silva, R., Aires, K., Santos, T., Abdala, K., Veras, R., Soares, A.: Automatic detection of motorcyclists without helmet. In: Computing Conference, Latin American (CLEI-2013), Naiguata, pp. 1–7, 7–11 Oct 2013
18. Doungmala, P., Klubsuwan, K.: Helmet wearing detection in Thailand using Haar like feature and circle hough transform on image processing. In: International Conference on Computer and Information Technology, pp. 611–614 (2016)
19. Dahiya, K., Singh, D., Mohan, C.K.: Automatic detection of bikeriders without helmet using surveillance videos in real-time. In: International Joint Conference on Neural Networks, pp. 3046–3051 (2016)
20. Waranusast, R., Bundon, N., Timtong, V., Tangnoi, C., Pattanathaburt, P.: Machine vision techniques for motorcycle safety helmet detection. In: 28th International Conference on Image and Vision Computing New Zealand, pp. 35–40 (2013)
21. Sutton, R.S.: Learning to predict by the methods of temporal differences. Mach. Learn. **3**(1), 9–44 (1988)
22. Stauffer, C., Grimson, W.: Adaptive background mixture models for real-time tracking. Int. Conf. Comput. Vis. Pattern Recogn. **2**, 246–252 (1999)
23. Otsu, N.: A threshold selection method from gray-level histograms. IEEE Trans. Syst. Man Cybern. **9**(1), 62–66 (1979)
24. Haralick, R.M.: A measure of circularity of digital features. IEEE Trans. Syst. Man Cybern. **4**, 394–396 (1974)

Affinity-Aware Synchronization in Work Stealing Run-Times for NUMA Multi-core Processors

B. Vikranth, Rajeev Wankar and C. Raghavendra Rao

Abstract Modern high-performance server systems are typically built as several multi-core chips put together in a single system. Each chip is connected to its local memory via an integrated memory controller (IMC) behaving as a node and hence the single machine behaving as non-uniform memory architecture (NUMA). Various user-level run-time systems adapt work stealing load balancing technique in multi-core processors. The work stealing run-times have to be aware of the topology of the processor on which they are running. Work stealing run-times on multi-core processors typically rely on lock-based synchronization to guarantee the coherency of shared mutable state. Synchronization constructs such as mutex locks, condition variables, and barriers are extensively used in implementation of these user-level work stealing run-times. The locality of these lock variables in multi-socket NUMA processors has considerable impact on the performance of these run-time systems. This paper studies the effect of locality of these synchronization constructs and proposes NUMA awareness to them. The proposed methodology is implemented using a source to source translator of OpenMP run-time, evaluated using OpenMP microbenchmark programs.

Keywords NUMA · Remote access · Work stealing · Stealing domain Synchronization

B. Vikranth (✉)
CVR College of Engineering, Hyderabad 501510, India
e-mail: b.vikranth@gmail.com

R. Wankar · C. Raghavendra Rao
SCIS, University of Hyderabad, Hyderabad 500046, India
e-mail: wankarcs@uohyd.ac.in

C. Raghavendra Rao
e-mail: crrcs@uohyd.ernet.in

© Springer Nature Singapore Pte Ltd. 2019   121
J. K. Mandal et al. (eds.), *Advanced Computing and Communication Technologies*,
Advances in Intelligent Systems and Computing 702,
https://doi.org/10.1007/978-981-13-0680-8_12

1   Introduction

Modern processors contain multiple cores on single chip keeping performance improvement as target. The memory technology is not progressing as that of CPU technology is progressing. As the number of cores is increasing, accessing the same memory using a common bus results in memory bottleneck. Though CPU operates at 3–4 GHz rate, memory cannot deliver the data to cache or CPU register file. The solution to this problem is introducing more than one memory controllers interfacing with DRAM. Many processors belonging to high-performance family such as Xeon E5-26xx series from Intel or Opteron 62xx series processors have two or more integrated memory controllers. The total number of CPUs is grouped into sockets. Each socket has a separate integrated memory controller (IMC) interacting with separate memory module to minimize overall memory latency. The processors (cores) on each socket are connected by interconnection network such as quick path interconnect (QPI) [1] from Intel or Hyper-Transport link from AMD [2]. The processors on the socket can access the memory locations via the IMC attached to the local socket at a faster rate. Nevertheless, when a processor on one socket tries to access a memory location attached to the other socket, the accessing speed is not same as that of the local memory access speed. Hence, these processors exhibit NUMA behavior. The fraction of the time taken to access a remote memory location to the time taken to access a local memory location is called NUMA ratio (R_{NUMA}) and is given by

$$R_{NUMA} = \frac{T_{remoteaccess}}{T_{localaccess}} \tag{1}$$

Operating systems have to understand the topology of the interconnection in NUMA environment. This information is helpful to allow the user applications to access the data from memory in such a way that the memory latencies are minimized. Since the Linux kernel uses the same system call *clone()* for creation of process and thread, thread and process are treated as same at kernel level using same `task_struct` object. Linux kernel follows first-touch policy [3]. The first-touch policy ensures the data locality to threads based on first write accesses to the memory location. It is not based on allocation affinity; i.e., physical memory for an object is not allocated on node when a thread requests heap memory allocation rather when actually thread initializes (write) the object for the first time. This is applicable to all types of dynamically allocated objects. Figure 1 depicts the scenario where the main thread whose affinity is on node 0 initializes a data structure. The threads, pinned on to the cores of other node, suffer from remote memory access latency to access the objects. Synchronization constructs such as spin lock, mutex, condition variables are not an exception for this first-touch policy. These primitive synchronization constructs are implemented using either busy waiting or blocking technique. If a single thread initializes all these objects, there is a possibility that in addition to busy wait overhead, they also suffer from remote memory access latency in multi-socket architectures.

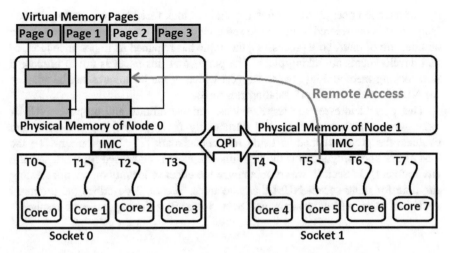

Fig. 1 All virtual pages mapped to physical memory of single node due to first-touch policy

1.1 User-Level Task Stealing Run-time Systems

Because of the creation overhead of process construct and thread constructs, many modern run-times are moved to user level. The parallel run-times such as Cilk [4], TBB [5] and OpenMP [6] operate at user level maintaining a pool of threads using the native thread library support of kernel. Instead of creating a thread for every parallel activity, these run-times provide a lighter construct than thread called task [7]. During initialization, the master thread of these run-times creates a pool of native threads called worker threads whose lifetime is till the end of parallel program execution [6]. After creation, these worker threads await user-created tasks, waiting on a condition variable associated with a mutex. These tasks are generally kept in a queue. The run-times of Cilk, TBB, and few implementations of OpenMP [8] extend the functionality of thread pool concept as work stealing pools where each worker thread maintains a separate double-ended queue [9, 10]. One end of the queue is accessible to worker thread to pop and execute the task body. The other end of the queue is accessed by idle workers having an empty task queue. The worker thread, which attempts to steal a task from other worker queue, is called a thief, and the worker thread from which a task is stolen is called a victim [11].

While running a parallel application, the affinity of threads can be controlled using explicit tools such as numactl [12] or likwid tools [13] or by setting environment variables such as KMP_AFFINITY from Intel compilers and OMP_PLACES in OpenMP 4.x specification [14]. Synchronization constructs such as lock variables, condition variables, and barriers are created and initialized by the master thread of work stealing run-time systems. If these are allocated by the master thread, according to first-touch policy, these constructs are allocated on the node where master thread is bound to. As a result, the worker threads which are bound to the cores of different nodes suffer from remote access latency on all primitive synchronization operations. Though the study of lock variables and synchronization techniques is considered to be an old research area, the recent developments in the architecture of multi-core processors, this topic has regained its importance [15, 16] and developments are taking place. The purpose of this paper is not to propose a new locking methodology, but to suggest which of the techniques can be adapted for NUMA multi-core work stealing run-times.

This paper addresses the issue of lock variable affinity and proposes NUMA extensions to `pthread` lock API particularly for work stealing run-times and evaluate the proposed method using microbenchmark [17, 18] programs. To the best of our knowledge, this is the first time the issue is addressed for work stealing environments. In Sect. 2, we try to analyze the effect of locality on a work stealing run-time for multi-socket NUMA architectures. Section 3 describes the proposed synchronization API extensions, and Sect. 4 describes the experimental results.

2 Effect of Locality on Synchronization Constructs

When a worker thread is created, it waits on a condition variable until the program adds tasks to its queue and the dispatcher sends these tasks to the worker's double-ended queue. The allocation of memory for the queue object, condition variable, and the associated mutex lock is allocated on the node where the master thread runs as per Linux's first-touch policy. But the worker thread may be bound to a core on different nodes (socket). To check if any tasks are available on its queue, the worker thread is engaged in busy waiting using operating system's atomic primitive such as CAS (compare and swap). If the worker thread attempts n times to read the condition variable's value which is allocated on local node, the busy waiting time can be given by (2). In (2) and (3), $T_{xxxLock}$ represents the time spent on *acquire_lock()* operation, $T_{localaccess}$ is the local memory access latency time and T_{CAS} is the time spent on *compare and swap* (CAS) operation at machine instruction level.

$$T_{localLock} = \left(\sum_{i=1}^{n} T_{localaccess} \right) + T_{CAS} \qquad (2)$$

If the condition variable and the associated mutex lock variables are stored on a remote node, the delay involved in busy waiting can even be greater than that of (2) because on every primitive CAS operation, to compare the lock variable value, it has to access remote memory location. In this scenario, the time spent in acquiring the lock is given by (3).

$$T_{remoteLock} = \left(\sum_{i=1}^{n} T_{remoteaccess} \right) + T_{CAS} \qquad (3)$$

Table 1 Memory latencies of dual socket Xeon E5-2620

Node	0	1
0	77.3 ns	124.7 ns
1	122.8 ns	75.0 ns

Table 2 Spin lock access delays local versus remote

Contending thread pinning	Average lock access time (ns)
On same socket	15.435
On different sockets	38.425

If synchronization objects are not NUMA-aware, they are susceptible to NUMA effects. These effects not only result performance mismatch between cores but also cause starvation or even live-lock under load. By the time, the lock is available and its status is known to the thread on other node, a local thread on same node where lock object is located may grab the lock. The effect of locality of spin locks is studied in [9]. These circumstances may cause the thread on remote node to starve. Optimizing the placement of shared lock objects across cores of different sockets minimizes NUMA effects. If shared lock is accessed by a group of threads on the single socket, and lock object is bound to remote memory, the threads suffer from remote access latencies to acquire lock. On the experimental set up of dual socket Intel's Xeon-E5 series machine running Linux kernel, when Memory Latency Checker program [19] was run, the results presented in Table 1 which yields average NUMA ratio $R_{NUMA} = 1.625$.

Hence, it can be observed from (2) and (3) that the additional delay involved in testing a remote condition variable and lock is about a factor of $(R_{NUMA} - 1) \sum_{i=1}^{n} T_{localaccess}$. This leads to a performance penalty of approximately R_{NUMA} times. The delays explained in the equations (1–3) are also applicable to worker thread barriers. If barrier variable is allocated on a different node, the worker threads have to experience the same delays while joining the worker threads. To evaluate the above theoretical concept, a simple spin lock-based program was run on dual socket Xeon E5-2620 Linux machine. Average starvation time of spin lock access by two threads sitting on same socket cores and cores of different sockets is presented in Table 2.

3 Thread Affinity Lock API for NUMA Multi-core Processors

The two possible improvements that can be incorporated as part of work stealing run-times in NUMA are as follows:

- Using NUMA-aware locks.
- Ensuring locality of the lock objects based on worker thread affinity.

Recently, lock cohorting [15] was developed as a generalized methodology for allowing NUMA awareness in locks. Lock cohorting approach is based on a combination of two locks: one lock used as a global lock and another used as local lock (there is one global lock for all nodes and one local lock per NUMA node) [16]. In the work stealing run-time, we group all worker threads pinned to the cores of single socket as stealing domain [20]. In a system with S sockets and C cores per socket, the run-time creates S stealing domains. If work stealing run-time is implemented using the proposed local locks along with lock cohorting, all C worker threads belonging to a single socket can depend on local lock for synchronization within the socket. When one of these C workers has to access a remote node for task stealing among S nodes, it can depend on global lock. Stealing domains put the best effort to minimize such remote stealing attempts. Usage of global lock is done only in rare case.

3.1 Affinity-Aware Worker Thread Implementation

Since pthread_xxx_init(), pthread_xxx_lock(), and pthread_xxx_unlock() API calls where xxx may take one of the possible option from mutex, spin types of POSIX standard locks are not aware of the first-touch awareness, we propose API with common syntax pthread_NUMA_xxx_lock() and pthread_NUMA_unlock(). Similarly, for condition variables, the proposed syntax is pthread_NUMA_ cond_init(), pthread_NUMA_cond_wait(), and pthread_NUMA_cond_signal(). These proposed API calls can collect the thread information to which node the thread is pinned to using libnuma API calls. The common steps in pseudocode form of the proposed API calls are presented in the following code snippet.

3.2 Implementation of Lock Cohorting in Stealing Domains

In the proposed system, the number of nodes and CPUs per node information is collected during initialization of run-time. This information can be obtained with the help of libnuma [12] API calls. Associated with each core (CPU), a new worker thread is created during initialization of the run-time. The group of worker threads that are pinned to the cores belonging to a single socket is called a stealing domain [20, 21]. The concept of stealing domain is introduced to put best efforts of allowing the worker threads to steal tasks only from worker queues belonging to the same node. It is an improvement to plain work stealing technique which selects the victim worker randomly. Stealing domains do not completely avoid remote node task stealing but minimizes the remote node steal attempts as best effort. By adapting affinity-aware synchronization, API described in Sect. 3 can be applied to these stealing domains. As a result, the worker threads belonging to same domain also operate on mutexes and condition variables belonging to same node.

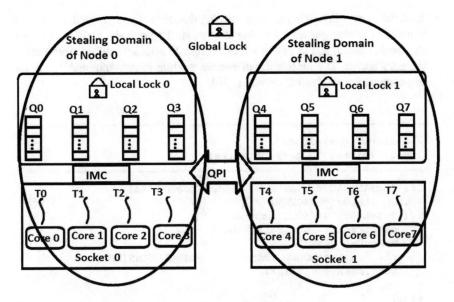

Fig. 2 Stealing domains and lock cohorting collaboration

The architecture of stealing domains along with lock cohorting is presented in Fig. 2. In this figure, worker threads are denoted and suffixed by T and respective worker queues by Q.

```
int thread_NUMA_xxx_init (pthread_xxx_t *var){
    tid = pthread_self ();
    cpu_id = sched_getcpu (tid);
    node = numa_node_of_cpu (cpu_id);
    var=numa_alloc_onnode(sizeof(*var),node);
}
```

Even the barriers can be restricted to the worker threads belonging to same stealing domain. The following two improvements are done to existing work stealing technique proposed in [20].

- The condition variable on which the worker thread waits is located on the same node where the worker thread is bound. Hence, the $T_{remoteaccess}$ component of (3) from the Section Introduction in each attempt is minimized.
- Whenever a worker thread finds no tasks in its queue, it becomes a thief and it attempts to steal a task from other queue belonging to the same domain. Before staling, the thief worker has to acquire a lock of the victim's queue. Since the queues and locks of workers are all located on the same node, the thief does not attempt any remote lock access.

- If all the queues belonging to same stealing domain are underloaded, an attempt is made to worker queues belonging to other stealing domain. Only in this case, the worker thread has to poll at a lock on remote node and suffers from remote memory access. The probability of remote stealing is very minimal due to the implementation of stealing domains [20].

```
Algorithm: WorkerRun
Input: Pointer to current worker
```

```
if( localTaskQueue.size == THRESHOLDMAXSIZE)
  this.status = VICTIM;
if( ! isEmpty ( localTaskQueue) ){
    popAtFront( localTaskQueue , task );
    execute( task );
    if( localTaskQueue.size == THRESHOLDMINSIZE)
    this.status = THIEF;
}
else{
    /* task stealing from local node */
    this.status = THIEF;
    if ( Global lock g is acquired by local node ){
      localStealingDomain.acquireLock();
      taskQueue = searchVictimQueue(localStealingDomain);
      popAtRear( taskQueue, task );
      localStealingDomain.releaseLock();
    }
    if ( task )
      execute(task);
    else{
      /* task stealing from remote node */
      if( Global lock g is acquired by remote node_i ){
        stealingDomain[i].acquireLock();
        taskQueue = searchVictimQueue(stealingDomain[i]);
        popAtRear( taskQueue, task);
        stealingDomain[i].releaseLock();
      }
      if ( task )
        execute(task);
    }
}
```
---

4   Results and Analysis

To evaluate the results of the proposed affinity-aware NUMA lock library, open-source OpenMP run-time OMPi [8] is considered as user-level work stealing infrastructure. OMPi is an implementation of OpenMP directives which supports processes, pthreads, and any other native threads. As part of this run-time, we replaced the lock initialization primitives in the file `othr.c` of `ee_pthreads` module. The function `othr_init_lock(othr_lock_t *lock, int type)` is modified using affinity-aware lock primitives and OpenMP microbench mark programs [17, 18] are run to evaluate the proposed strategy.

OpenMP microbenchmark is a set of programs to evaluate the implementation overheads of synchronization, parallel for, arrays loop, and scheduling constructs offered by OpenMP specification. The set of programs include array-based programs of various sizes, scheduling benchmark whose purpose is to evaluate the scheduling overhead in OpenMP implementation. `syncbench` program of the benchmark suits was considered to evaluate the NUMA-aware synchronization constructs such as locks and mutexes. The experimental environment is a dual socket Xeon E5 2620 series processor running Linux kernel 3.16 version. After running the experiments for 10 times, average overhead of various synchronization constructs is collected from the output of the benchmark. The results are presented in Table 3.

- The overhead of OpenMP constructs such as `parallel` and `parallel for` is considerable and is due to the worker threads waiting on condition variable located on remote node for the chunks of work.
- `barrier` construct is also a kind of lock where all the worker threads wait for others to finish. Locality of barrier object contributes to the difference in performance.
- `lock/unlock` and `atomic` constructs of OpenMP are internally translated to native mutex locks and the difference in performance is due to the locality of lock variable across the nodes.

Table 3 Comparison of synchronization constructs implementation overheads

OpenMP construct	Implementation overhead in ms	
	NUMA oblivious synchronization	Affinity-aware synchronization
parallel	10.821	8.661
parallel for	6.893	6.090
barrier	2.219	2.092
critical	0.252	0.244
lock/unlock	0.291	0.267
atomic	0.365	0.158
reduction	8.805	6.839

Table 4 Comparison of remote data volumes accessed by benchmark

NUMA oblivious synchronization	Affinity-aware synchronization
0.0772 GB	0.0420 GB

Fig. 3 Comparison of implementation overheads

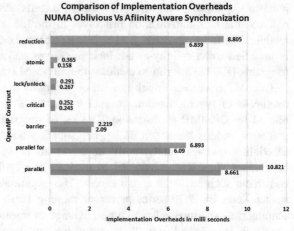

- `reduction` overhead is due to the critical section code placed at the end of parallel for in its expanded form uses again locks for protection of reduction variable. The sync benchmark contains many parallel for with reduction clause. Hence, the difference of overheads is more.

To crosscheck whether the overhead observed is occurring due to remote memory access delays or for some other reason, `likwid` tools are used to measure performance counter values related to uncore NUMA events. The sum of remote data volumes are presented in Table 4. It can be easily concluded that the excess remote data volume in the default approach is the cause of additional overheads observed in Table 3 (Fig. 3).

5   Conclusion

This paper is an effort put to analyze the importance of locality of synchronization objects in implementation of work stealing-based environment in NUMA multi-core processors. The analysis is helpful in adapting existing work stealing-based run-times to NUMA multi-core which use native thread library and synchronization constructs. If NUMA-aware locks can be incorporated in implementation of these run-times, the performance of run-time and target application can be improved. These locality-aware constructs can be implemented at run-time layer and do not affect the source code of the application.

References

1. Ziakas, D.: Intel quick path interconnect architectural features supporting scalable system architectures. In IEEE 18th Annual Symposium on High Performance Interconnects (HOTI), pp. 1–6 (2010)
2. Hughes, B., Conway, P.: The AMD Opteron northbridge architecture. In IEEE Micro **27**(2) (2007)
3. Majo, Z., Gross, T.: Memory system performance in a NUMA multicore multiprocessor. In Proceedings of the 4th Annual International Conference on Systems and Storage, p. 12. ACM, 30 May 2011
4. Joerg, C.F., Kuszmaul, B.C., Leiserson, C.E., Randall, K.H., Zhou, Y., Blumofe, R.D.: Cilk: an efficient multithreaded runtime system. J. Parallel Distrib. Comput. **25**(37), 55–69 (1996)
5. Pheatt, C.: Intel® threading building blocks. J. Comput. Sci. Coll. **23**(4), 298–299 (2008)
6. Terboven, C., Wong, M., an Mey, D., Eichenberger, A.E.: The design of OpenMP thread affinity. In OpenMP in a Heterogeneous World, pp. 15–28 (2012)
7. Hadjidoukas, P.E., Agathos, S.N., Dimakopoulos, V.V.: Design and implementation of openmp tasks in the ompi compiler. In 15th Panhellenic Conference on Informatics (PCI), pp. 265–269. IEEE, 30 Sept 2011
8. http://paragroup.cse.uoi.gr/wpsite/software/ompi/
9. Al Bahra, S.: Nonblocking algorithms and scalable multicore programming. Queue **11**(5), 40 (2013)
10. Lev, Y., Chase, D.: Dynamic circular work-stealing deque. In Proceedings of the Seventeenth Annual ACM Symposium on Parallelism in Algorithms and Architectures ACM, pp. 21–28, 18 July 2005
11. Leiserson, C.E., Blumofe, R.D.: Scheduling multithreaded computations by work stealing. J. ACM (JACM) **46**(5), 720–748 (1999)
12. Kleen, A.: A NUMA API for Linux. Novel Inc. (2005)
13. Hager, G., Wellein, G., Meier, M., Treibig, J.: LIKWID: lightweight performance tools. In Proceedings of the 2011 Companion on High Performance Computing Networking, Storage and Analysis Companion, pp. 29–30. ACM (2011)
14. OpenMP AR. OpenMP 4.0 specification. June 2013
15. Marathe, V.J., Shavit, N., Dice, D.: Lock cohorting: a general technique for designing NUMA locks. ACM Trans. Parallel Comput. **1**(2), 13 (2015)
16. Marathe, V.J., Shavit, N., Dice, D.: Lock cohorting: a general technique for designing NUMA locks. In ACM SIGPLAN Notices, vol. 47, no. 8, pp. 247–256. ACM (2012)
17. O'Neill, D., Bull, J.M.: A microbenchmark suite for OpenMP 2.0. In ICPP, ACM SIGARCH Computer Architecture News, vol. 29, no. 5, pp. 41–8, 1 Dec 2001
18. Reid, F. McDonnell, N., Bull J.M.: A microbenchmark suite for openmp tasks. In International Workshop on OpenMP, pp. 271–274. Springer, Berlin, Heidelberg, 11 June 2012
19. https://software.intel.com/en-us/forums/software-tuningperformance-optimization-platform-monitoring/topic/600141
20. Wanker, R., Raghavendra Rao, C., Vikranth, B.: Topology aware task stealing for on-chip NUMA multi-core processors. In Procedia Computer Science (ICCS'13), pp. 379–388 (2013)
21. Wanker, R., Raghavendra Rao, C., Vikranth, B.: Effective task binding in work stealing runtimes for NUMA multi-core processors. IJCSE, **8**(4), pp. 189–196 (2017)

Back-Propagation Neural Network Versus Logistic Regression in Heart Disease Classification

Shrinivas D. Desai, Shantala Giraddi, Prashant Narayankar,
Neha R. Pudakalakatti and Shreya Sulegaon

Abstract Globally, cardiovascular (heart) diseases are the major cause of death. About 80% of deaths are reported in developing countries. Looking at the trend and lifestyle, one can predict that by 2030 around 23.6 million people may die due to heart disease (mainly from heart attacks and strokes). Each and every healthcare unit generates enormous heart disease data which unfortunately are not "mined" to discover pattern and knowledge for effective decision making. Practical knowledge by domain experts plays vital role. However, there is a need for effective analysis tools to discover unknown relationships and trends in data. Objective of this paper is to assess the accuracy of classification model for the prediction of heart disease for Cleveland dataset. A comparative study of parametric and nonparametric approach in classifying heart disease is presented. Two classification models, back-propagation neural network (BPNN) and logistic regression (LR), are used for the study. The developed classification model will assist domain experts to take effective diagnostic decision. 10-fold cross validation method is used to measure the unbiased estimate of these classification models.

Keywords Heart disease · Prediction · Artificial neural network
BPNN · Accuracy

1 Introduction

In US, approximately 8 lakh death is reported due to cardiovascular disease (CVD) in 2017. On a global level, this CVD accounts for 31% (17.7 million) of all deaths. CVD, cardiomyopathy, and coronary heart disease are various forms of heart diseases which may result due to various factors such as blood pressure, obesity, smoking. Accurate diagnosis of heart disease at affordable price is a great

S. D. Desai (✉) · S. Giraddi · P. Narayankar · N. R. Pudakalakatti · S. Sulegaon
School of Computer Science and Engineering, KLE Technological University,
Hubballi, India
e-mail: shree.desai07@gmail.com

© Springer Nature Singapore Pte Ltd. 2019
J. K. Mandal et al. (eds.), *Advanced Computing and Communication Technologies*,
Advances in Intelligent Systems and Computing 702,
https://doi.org/10.1007/978-981-13-0680-8_13

challenge [1]. High-quality clinical decisions lead to efficient diagnosis, while poor-quality decisions lead to improper diagnosis of the patient. The need of the hour is high-quality clinical decisions at low cost. This can be achieved by applying advanced data mining techniques which helps in discovering hidden patterns in healthcare datasets [2, 3].

The artificial neural network (ANN) and LR, the machine learning algorithms are used in healthcare industry to minimize the diagnosis cost [4, 5]. Both parametric and nonparametric models are used to solve classification problems. Classification accuracy in ANN can be enhanced by rearranging the weights of neurons [6]. ANN and LR with decision support play a very important role in improving the quality of service in health care or in medical centers. Thus, ANN model-based diagnoses have significantly helped health care by providing effective diagnosis [7].

1.1   Artificial Neural Network Architecture (ANN)

Artificial neural network can be divided into three layers, namely input layer, hidden layer, and output layer [8]. Figure 1 shows general structure of ANN [9].

- Input layer: This layer receives data from the external environment. These inputs are usually normalized which results in better numerical precision for the mathematical operations performed by the network.
- Hidden layers: These layers are composed of neurons which are responsible for extracting patterns associated with the data that is fed to the neural network.
- Output layer: The results of the processing performed by the neurons in the previous layers are produced by these neurons.

1.1.1   Number of Hidden Neurons

Selection of number of hidden neurons is very important for getting good results. Inappropriate number of hidden neuron causes either overfitting or underfitting. If the number of neurons is less as compared to the complexity of the problem data,

Fig. 1 ANN architecture. *Courtesy* A Medium Corporation

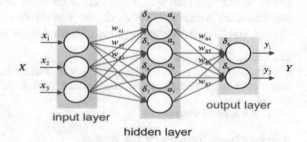

then "Under fitting" may occur. If the number of neurons is more for the complexity of problem, then "Over fitting" may occur. Till now, there is no exact formula for computing the number of hidden neurons. Most of the times, it is trial and error method.

However, there are several rule-of-thumbs for choosing the number of hidden neurons.

- The number of hidden layer neurons is 2/3 of the size of the input layer [10]
- The number of hidden layer neurons should be less than twice of the number of neurons in input layer [3]
- The size of the hidden layer neurons is between the input layer size and the output layer size [11].

1.1.2 Back-Propagation Algorithm

Back-propagation algorithm is a popular supervised learning model [11]. The BPNN adopts the principle of multi-layer feed-forward network which models a function by performing error calculation of each neuron [12]. BPNN keeps on updating its weight after processing the group of data until error values are within the threshold [10, 13].

1.2 Logistic Regression

LR is a parametric model where the categorical variables are dependent [14]. This parametric model is used for classifying categorical values, whose values depend on the success of probability. Figure 2 depicts the LR model. It involves fitting an equation of the form:

Fig. 2 Logistic regression model

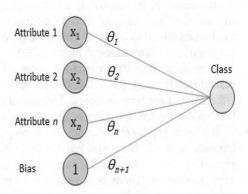

$$i = 1 \Big/ \left(1 + e^{((x0 + x1)*j)}\right) \tag{1}$$

i = input values, j = output values, $x0$ = bias term, $x1$ = coefficient for single input value (i).

2 Related Work

Vikas et al. (2013) proposed data mining approach to detect heart disease using Naive Bayes, decision tree, and bagging algorithm. In this paper, a systematic comparison is carried out to between bagging and non-bagging algorithms and accuracy was found to be 85.03% on bagging algorithm [15].

Kirmani [16] proposed a prediction model using multi-layer perceptron algorithm. Input dataset with 13 attributes and 303 records was made as input to the algorithm. Out of which, 245 records were accurately diagnosed with accuracy of 80.85%.

Abdar et al. [17] proposed a comparative study of various data mining approaches for prediction of heart disease. Input datasets with 13 attributes and 270 records were made as input to the several algorithms like C5.0, support vector machine (SVM), k-nearest neighbor (k-NN), and neural networks. Decision tree algorithm-C5.0 model, found to produce highest accuracy of 93.02%.

Gandhi and Singh [14] proposed a comparative study on classification model in predicting heart disease using decision tree, neural network, and Naive Bayes. Through the comparative study, a better hybrid model could be built in predicting heart disease.

Sayed and Halkarnikar [18] proposed a genetic neural-based algorithm for prediction of heart disease. Input dataset with more than 11 attributes was made as input to the algorithm. The model was implemented in MATLAB environment, and accuracy was found to be 89% [16, 18].

Sayad et al. [19] proposed a neural network-based approach for diagnosis of heart disease. In this approach, a multi-layer perceptron with BPNN model is built to train the input datasets with 13 attributes and 70 records. The accuracy of the model was found to be 94%. This paper is considered as a benchmark, and results are compared with proposed method [19].

Giraddi et al. [20] proposed a system based on BPNN for the detection of diabetic retinopathy (DR). DR is a complication in which vision of a person gets affected by long-term diabetes. The author explored various architectures to find best model for the detection of diabetic retinopathy.

Research gap noticed through the survey is: It is observed that various machine learning methods are employed for identifying accurate classification model for predicting heart disease. Thirteen attributes have become de facto dataset for experimentation. There is a need for comparing parametric and nonparametric approach in classifying toward predicting heart disease.

3 Data Description

Cleveland heart disease dataset has been used for the study [21]. Database consists of 303 instances with 13 attributes each, out of which 270 instances are used for the study. The database is prepared by V.A. medical center, long beach, and Cleveland clinic foundation. Description of attributes used for classification is presented in Table 1. Dataset is used for academic research with due permission request to concerned authority.

Table 1 Dataset description

Attribute number	Attributes	Description	Range
1	Age	Age in year	Continuous
2	Sex	(1 = male, 0 = female)	Discrete
3	Cp	Value 1: typical angina Value 2: atypical angina Value 3: non-angina pain Value 4: asymptomatic	Discrete
4	Trestbps	Resting blood pressure (in mm Hg on admission to hospital)	Continuous
5	Chol	Serum cholesterol in mg/dl	Continuous
6	Fbs	(Fasting blood sugar > 120 mg/dl) (1 = true, 0 = false)	Discrete
7	Restecg	Resting electrocardiographic results Value 0: normal Value 1: having ST-T wave abnormality (T wave inversions and/or ST Elevation or depression of > 0.05 mV) Value 2: showing probable or definite left ventricular hypertrophy by Estes criteria	Discrete
8	Thalach	Maximum heart rate achieved	Continuous
9	Exang	Exercise induced angina (1 = yes, 0 = no)	Discrete
10	Oldspeak	ST depression induced by exercise relative to rest	Continuous
11	Slope	The slope of the peak exercise ST segment Value 1: up sloping Value 2: flat Value 3: down sloping	Discrete
12	Ca	Number of major vessels (0–3) colored by fluoroscopy	Continuous
13	Thal	Normal fixed defect, reversible defect	Discrete

4   Methodology

The implementation of models is carried out using Anaconda Python language.

- Normalization: The dataset has been normalized to have a mean of zero and variance 1.
- Hidden layers and hidden neurons: The number of neurons is decided based on various thumbs-of-rules given in Sect. 1.
- Training Parameters: The network has been trained using gradient descent algorithm in batch mode (trained). The number of epochs is set to 200. Learning rate is set to 0.05.

4.1   Performance Evaluation Parameters

Performance of proposed classification models is assessed by well-known parameters such as accuracy, receiver operating curve (ROC), and confusion matrix.

4.2   Hypotheses Testing

In this section, different hypothesis testing such as *T*-test, *Z*-test, *F*-test, and chi-square test which are used for testing of null hypothesis is given below. The results of these testing are used to optimize the input parameters.

Z-test

$$Z_{\text{data}} = \frac{p1 - p2}{\sqrt{P\text{pooled} * (1 - P\text{pooled})\left(\frac{1}{n^2} + \frac{1}{n^2}\right)}} \tag{2}$$

where

$$P\text{pooled} = \frac{x1 + x2}{n1 + n2}$$

T-test

$$t_{\text{data}} = \frac{\bar{x}1 - \bar{x}2}{\sqrt{\frac{S1^2}{n1} + \frac{S2^2}{n2}}} \tag{3}$$

F-test

$$\text{MSTR} = \frac{\sum n1(x1 - \bar{x})^2}{k - 1}, \text{MSE} = \frac{\sum (n1 - 1)S1^2}{n_t - k} F_{\text{data}} = \frac{\text{MSTR}}{\text{MSE}} \quad (4)$$

(annova test)

Chi-squre-test

$$\text{Chi}^z_{\text{data}} = \sum \frac{(O - E)^2}{E} \quad (5)$$

(a) x1, x2—elements belonging to different group b. O—observed frequency c. E —expected frequency

5 Results and Discussion

5.1 Back-Propagation Neural Network (BPNN) Versus Linear Regression

The proposed heart disease classification is carried out with different architecture of ANN using back-propagation algorithm. The results of 10-fold cross validation for these architectures are given in Table 2. Table 3 shows the confusion matrices obtained for 27 samples. Figure 3 shows the classification accuracies obtained. Figure 4 shows the ROC curve for BPNN and that of Fig. 5 is ROC for linear regression.

Table 2 Percentage of MSE accurcay with BPNN architecture

Architecture	MSE	Accuracy (%)
13-6-6-1	0.2962	72.72
13-6-6-6-1	0.1851	85.07
13-6-6-6-6-1	0.2222	79.25
13-9-9-9-1	0.1851	78.51
13-12-12-1	0.2222	78.88

Table 3 Confusion matrices for various architectures

Architecture	PT	FP	TN	FN	SN (%)	SP (%)
13-6-6-1	0.3407	0.437	0.118	0.103	76.66	70.66
13-6-6-6-1	0.3444	0.4518	0.103	0.1	77.5	71.56
13-6-6-6-6-1	0.33	0.455	0.1	0.107	75.83	69.21
13-9-9-9-1	0.329	0.459	0.962	0.114	76.66	71.37
13-12-12-1	0.318	0.466	0.88	0.318	74.16	70.70

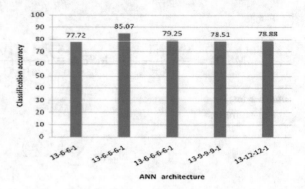

Fig. 3 Classification accuracies with various architectures

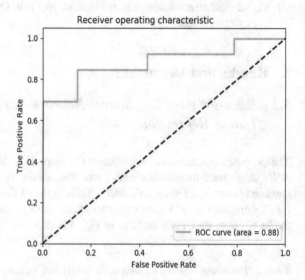

Fig. 4 ROC for BPNN

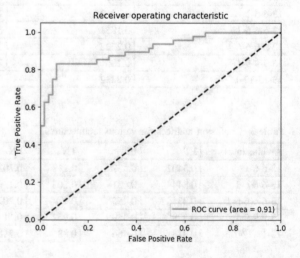

Fig. 5 ROC for linear regression

5.2 Hypothesis Testing and Analysis

It is very important to study each and every attribute and its influence on the classification model. Hence, following null hypothesis testing is defined and validated using *P*-test.

Case 1: *P*-test is conducted to test to check is there any significant difference in mean cholesterol of population significantly vary with respect to heart disease.

Inference: The observed *p*-value > 0.025 (95% CI); hence, it fails to reject the null hypothesis. There is no significant difference in cholesterol, with respect to heart disease. This indicates cholesterol is not one of the risk factors related to heart disease.

Case 2: Is there any significant difference in mean cholesterol of population significantly vary with respect to heart disease.

Inference: The observed *p*-value < 0.025 (95% CI); hence, we have evidence to reject the null hypothesis. There is a significant difference in blood pressure, with respect to heart disease. This indicates that prolonged high BP is one of the risk factors to get into a heart disease.

Case 3: Is there any significant difference in mean age of population significantly vary with respect to heart disease.

Inference 3: The observed *p*-value < 0.025 (95% CI); hence, we have evidence to reject the null hypothesis. There is a significant difference in age, with respect to heart disease. This indicates that age is one of the risk factors to get a heart disease.

Case 4: Is there any significant difference in mean age of population significantly vary with respect to heart disease.

Inference: The observed *p*-value < 0.025 (95% CI); hence, it fails reject the null hypothesis. There is a significant difference in heart rate, with respect to heart disease. This indicates that heart rate is one of the risk factors to get into a heart disease

Table 4 presents the *p*-value obtained for all the above four cases. In order to study the effect of more than two input attributes and their impact on the heart disease classification, Z-test, F-test, and chi-square tests are carried out.

Case 5: Whether exercise-doing people have a less risk of heart disease.

Inference: The observed *p*-value < 0.025 (95% CI); hence, it rejects the null hypothesis. There is a significant difference in exercise, with respect to heart disease. Hence, exercise-doing people have less risk of heart disease when compared to non-exercise people.

Table 4 *P*-test results

Number of case	Heart disease	*p*_value
Case 1	Mean cholesterol	0.05273
Case 2	Mean BP	0.01056
Case 3	Mean age	0.00044
Case 4	Mean heart rate	7.1103e−13

Table 5 Z-test results

Number of cases	Heart disease	Z_value	P-value
Case 5	Exercise	−10.365	0.0
Case 6	Gender	−11.354	0.0

Case 6: Whether the proportion of heart disease varies with respect to gender (male/female).

Inference: The observed p-value < 0.025 (95% CI); hence, it rejects the null hypothesis. There is a significant difference in gender, with respect to heart disease. Hence, based on the gender, there is a significant difference with respect to heart disease. Table 5 presents the z-value and p-value obtained for both the case.

Case 7: To check whether there is any significant difference with mean cholesterol with respect to vessels.

Inference: The observed p-value > 0.025 (95% CI); hence, it fails to reject the null hypothesis. That is there is no significant difference in mean cholesterol level with respect to vessel blockage.

Case 8: To check whether there is any significant difference of mean age with respect to vessels.

Inference: The observed p-value < 0.025 (95% CI); hence, it rejects the null hypothesis. That is there is a significant difference in age with respect to vessel blockage.

Table 6 presents the calculated p-value and F-value with respect to MSE and MSTR

Case 9: To check whether there is any significant difference between the genders with respect to vessels blockage.

Inference: The observed p-value > 0.025 (95% CI); hence, it fails to reject the null hypothesis. There is no significant difference in gender with respect to vessels. Which indicates gender is not one of the risk factors related to vessels.

Table 7 presents the calculated chi-square test and its corresponding p-value.

Results of proposed BPNN-based heart disease classification are compared with the recent literature [16], where multi-layer perceptron neural network is employed for the same classification and is presented in Table 8.

It is worth noting that accuracy of BPNN-based classification model is approximately 5% more as compared to the recent literature [16].

Table 6 F-test results

Number of case	Vessels	MSTR	MSE	$F =$ MSTR/MSE	P-value
Case 7	Cholesterol	4162.7	2654.6	1.5681	0.1974
Case 8	Age	1057.0	71.98	14.683	7e−09

Table 7 Chi-square results

Number of cases	Chi-square test	P-value
Case 9	6.4626	0.039506

Table 8 Results of proposed methods with other related works

Methodology	Datasets and records	Accuracy (%)	Author
Multi-layer perceptron BPNN	13 attributes 303 records	80.85	[1, 16]
BPNN	13 attributes 270 records	85.07	Proposed
LR	13 attributes 270 records	92.58	Proposed

6 Conclusions and Future Scope

Two objectives of this paper are identifying the best classification model among parametric and nonparametric for effective heart disease prediction and optimizing the number input attributes for classification model which were achieved by various systematic and engineering approaches. Experiments are conducted to find the best classifier for predicting the existence of heart disease. Input dataset having 13 fields and 270 records collected from Cleveland heart disease dataset is employed to validate the system. Among 13 fields, only 11 fields are most influencing and contributing parameters. This is achieved by testing various null hypotheses and cross-correlation matrix. The output variable is categorical in nature and the parametric based algorithms outputs are found to be more effective. Accuracy of 85.074 and 92.58% is recorded for BNN (nonparametric) and LR (parametric) models, respectively. With 10-fold cross validation and ROC analysis, area of 0.88 and 0.91 is recorded for BNN and LR, respectively. The empirical results show that with parametric model-LR, we can test effectively any new case of heart disease patient with optimal 11 parameters. At any moment of time, the model proposed cannot be replacement for clinical experts. It only complements the decision of clinician for taking better diagnostic decisions.

The futuristic scope of this project is to collect the dataset from local regional healthcare unit, along with additional input attributes such as smoking/alcoholic habits, food habit, and genetic conditions. Apply well-known parametric classification model and even designing hybrid model for effective heart disease prediction.

Acknowledgements Authors acknowledge "UCI Machine Learning Repository" for heart disease dataset. This work is partially supported by KLE Tech University under "Capacity Building Projects" grants. Authors acknowledge KLE Society and KLE Tech University, Hubli, for providing funding and support.

References

1. Dewan, A., Sharma, M.: Prediction of heart disease using a hybrid technique in data mining classification. IEEE (2015)
2. Palaniappan, S., Awang, R.: Intelligent heart disease prediction system using data mining techniques. In: IEEE (2008)

3. Berry, M.J.A., Linoff, G.: Data Mining Techniques. Wiley, New York (1997)
4. Bhatla, N., Jyoti, K.: An analysis of heart disease prediction using different data mining techniques. Int. J. Eng. Res. Technol. **1**(8) (2012). ISSN: 2278-0181
5. http://akri.org/intelligence/machine-memory.html
6. Vijayarani, S., Sudha, S.: Comparative analysis of classification function techniques for heart disease prediction. Int. J. Innov. Res. Comput. Commun. Eng. **1**(3) (2013)
7. Murthy, H.N., Meenakshi, M.: ANN model to predict coronary heart disease based on risk factors. Bonfring Int. J. Man Mach. Interface **3**(2) (2013)
8. Artificial Neural Networks: https://doi.org/10.1007/978-3-319-43162-8_2
9. Patel, A.R., Joshi, M.M.: Heart diseases diagnosis using neural network. In: IEEE 31661 (2013)
10. Boger, Z., Guterman, H.: Knowledge extraction from artificial neural network models. In: IEEE Systems, Man, and Cybernetics Conference, Orlando, FL, USA (1997)
11. Blum, A.: Neural networks in C++. Wiley, New York (1992)
12. Sonawane, J.S., Patil, D.R.: Prediction of heart disease using multilayer perceptron neural network. In: IEEE (2014)
13. Yuan, J., Yu, S.: Privacy preserving back-propagation neural network learning made practical with cloud computing. IEEE Trans. Parallel Distrib. Syst. **25**(1) (2014)
14. Gandhi, M., Singh, S.N.: Predictions in heart disease using techniques of data mining. In: International Conference on Futuristic Trends on Computational Analysis and Knowledge Management (ABLAZE). IEEE (2015)
15. Chaurasia, V., Pal, S.: Data mining approach to detect heart diseases. (2014)
16. Kirmani, M.M.: Heart disease prediction using multilayer perceptron algorithm. Int. J. **8**(5) (2017)
17. Abdar, M., Zomorodi-Moghadam, M., Das, R., Ting, I.H.: Performance analysis of classification algorithms on early detection of liver disease. Expert Syst. Appl. **67**, 239–251 (2017)
18. Amin, S.U., Agarwal, K., Beg, R.: Genetic neural network based data mining in prediction of heart disease using risk factors. In: IEEE Conference on Information and Communication Technologies (2013)
19. Sayad, A.T., Halkarnikar, P.P.: Diagnosis of heart disease using neural network approach. In: Proceedings of IRF International Conference (2014)
20. Giraddi, S., Pujari, J., Seeri, S.: Role of GLCM Features in Identifying Abnormalities in the Retinal Images. IJIGSP **7**(6), 45–51 (2015). https://doi.org/10.5815/ijigsp.2015.06.06
21. Dua, D., Taniskidou, E.K.: UCI machine learning repository [http://archive.ics.uci.edu/ml]. University of California, School of Information and Computer Science, Irvine, CA (2017)

Part II
Communication Technologies

Wormhole Attack in Wireless Sensor Networks: A Critical Review

Nishigandha Dutta and Moirangthem Marjit Singh

Abstract A Wireless sensor network (WSN) consisting of spatially distributed autonomous devices using sensors is prone to numerous kinds of threats and attacks due to various reasons such as unattended nature of deployment in an un-trusted environment, limited network resources, easy network access as well as range of radio transmission. Wormhole attack is one such attack where an intruder establishes a low latency link between two sensor nodes so as to misguide the nodes and exhaust network resources by gaining access to sensitive information. This paper outlines the wormhole attack in WSN and provides a critical review on the techniques to deal with it. Various wormhole detection techniques and prevention mechanisms that can be used to counter the wormhole attack in WSN are compared and analyzed briefly in this paper.

Keywords Wireless sensor network · Wormhole attack · Security attacks

1 Introduction

As wireless sensor network, the emerging technology, could provide flexible infrastructure in many real-world applications, e.g., health care, industrial automation, surveillance, and defense, security of such applications is a major concern. Most of these applications carry sensitive information too and hence require secured communication to protect against possible attacks.

A large number of sensor nodes constitute the wireless sensor network which are capable of sensing, transmitting, and processing different environmental parameters but are tiny in size and have limited bandwidth, power, and energy. Due to such

N. Dutta · M. M. Singh (✉)
Department of Computer Science& Engineering, North Eastern Regional
Institute of Science & Technology, Nirjuli 791109, India
e-mail: marjitm@gmail.com

N. Dutta
e-mail: nishidtt9@gmail.com

© Springer Nature Singapore Pte Ltd. 2019

147

J. K. Mandal et al. (eds.), *Advanced Computing and Communication Technologies*,
Advances in Intelligent Systems and Computing 702,
https://doi.org/10.1007/978-981-13-0680-8_14

limitations, the sensor nodes as well as the wireless sensor network itself are susceptible to several kinds of attacks carried out by an attacker to access and manipulate sensitive information as well as distort and damage the whole network. Since wireless sensor network carries sensitive data transmitted in a broadcast nature and also the deployment of sensor nodes is in an unattended and hostile environment, therefore implementing security measures in such type of network is necessary. Also, due to various constraints on the sensor nodes, the security measures taken must be different from those applied in conventional wired or wireless networks [1].

2   Wormhole Attack in WSN

Wormhole attack is one of the severe kinds of attack in a wireless sensor network which requires at least two or more adversaries in order to launch an attack on the network. The malicious nodes present in the network establish a communication link (tunnel) between them, and the tunnel is used to transport data from one place to another in the network [2]. The tunnels are established either by using sophisticated antennas or wired links placed in places which are physically remote in nature [1]. The wormhole attack is considered as a dreadful attack due to its ability to instigate other critical kinds of attacks such as black hole attack, sinkhole attack, and grayhole attack [3]. Also, wormhole attack can be performed in spite of implementation of encryption and authentication methods in the network. Thus, detection of this attack and defending against it have become a necessity in WSN.

For launching a wormhole attack, the attacker first establishes a high bandwidth, low latency link (tunnel) known as wormhole link. This high-speed off-channel link is constructed between two malicious nodes located at remote locations to each other in the sensor network. Thus, in a wormhole attack, attackers form pairs of wormhole links to transport data packets from end to the other [4, 3]. Once these tunnels are created, they are presented to the base station of the sensor network as being high-quality paths by the malicious nodes. Thus, these malicious nodes are established in such a way that they are presented as neighbors to all the sensor nodes of the network. The legitimate nodes adopt these wormhole links as the shortest route of communication, making data access possible for the attackers. The attackers then collect data packets at one end of the tunnel and replay them at the other end of it [5]. The working of the wormhole attack is shown in Fig. 1. The malicious nodes A and B are far away from each other and require several hops for a packet to be transferred from node A to B and vice versa. But due to the wormhole link formed between them, the attacker concludes A and B as neighbors to rest of the nodes in the sensor network, and with the help of routing messages, it eventually disrupts communication in the network by dropping data packets [3]. Let us consider that node C and node Q want to exchange data packets in the network. So node C broadcasts the route request in the network. Upon receiving the request, node A replays this message to B, who rebroadcasts the request to its neighboring

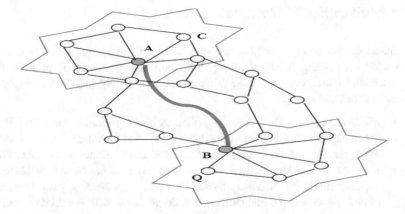

Fig. 1 Example of wormhole attack using a single wormhole link

nodes. Once node Q receives the route request through B, it assumes C to be its direct neighbor and sends reply to the route request. Node B then captures the reply and forwards it to node A, who transfers it to node C using similar procedure. Ultimately, both node C and Q will have the misconception of being two hops away from each other, and the entire communication will take place through the tunnel starting at node A and ending at node B [6].

Out of all the attacks on a wireless sensor network, wormhole attack is a brutal one as it exploits confidentiality, availability, and overall security of the sensor network. The drastic impact on the network due to the instigation of the wormhole attack is discussed below:

- *Manipulation of routing protocols*: The wormhole attack significantly deteriorates the functioning of network protocols by taking control over the routing traffic of the sensor network. Once an intruder gets access to routing information, it effects neighborhood discovery by creating fake list of neighbors and disrupts network functioning by dropping data packets [7, 1].
- *Layout for several other attacks*: The wormhole attack serves as a layout for several other harmful attacks which are mentioned below:
  1. *Rushing attack*: An attacker in a wormhole attack is able to attract traffic from the neighboring nodes if the wormhole link being established has a faster transmission rate. This in turn results in rushing attack where all the packets in the network adapt this fast wormhole link as their transmission channel, leading to an increase in the average attack success rate [8].
  2. *Sinkhole attack*: While performing the wormhole attack, the malicious nodes present at the two ends of the tunnel gathers required information so as to launch sinkhole attack in the sensor network [1].
  3. *Selective forwarding attack*: Wormhole attack can also further lead to selective forwarding attack as an intruder can block and drop certain packets through the two malicious nodes that form the wormhole tunnel [1].

3   Classification of Wormhole Attack

A. Based on the absence or presence of identities of malicious nodes and their packet forwarding behavior during tunneling and replaying of packets, the wormhole attack occurs in two different modes, namely hidden and exposed modes, respectively [2].

   • *Hidden Mode*: In this mode, the malicious nodes present in the network do not manipulate the content of the data packets and the AODV packet header while transferring the packets from one end of the tunnel to the other end.
   • *Exposed Mode*: The attacker manipulates the contents of data packets in exposed mode by including its identity while transferring packets in the tunnel. However, the malicious nodes do not mess with the AODV packet header, and it remains unaltered [7].

B. Based upon the implementation method used, the wormhole attack can be classified into the following categories:

   • *Out-of-band channel*: In this mode, the malicious nodes are connected by a high-quality and high-bandwidth out-of-band channel in order to launch the wormhole attack. The tunnel thus established can be constructed either by using a wired link or a directional wireless link. This mode requires specialized hardware for implementation making it more difficult to launch compared to other methods such as encapsulation [6, 9].
   • *Packet Encapsulation*: Implementation of wormhole attack becomes easy with the use of encapsulation method as it does not require any special hardware, plus the ends of the wormhole link do not contain any cryptographic data. In this mode, the two malicious nodes contain several sensor nodes in between and the data packets flowing between them are encapsulated. Due to this encapsulation, the actual hop count of the nodes does not get incremented and the data packets are transformed to their original form at the end point of the tunnel [2, 9].
   • *Packet Relay*: Wormhole attack can be launched by packet relay method using one or more malicious nodes. In order to create fake list of neighbors, the attacker convinces two remote sensor nodes that they are neighbors by replaying packets between them [9].
   • *High Power Transmission*: This type of wormhole attack is launched by a single malicious node with a high power transmission capability. The malicious node transfers data packets to other sensor nodes from a remote distance at high power so as to establish itself into the path between the source and destination [9].

C. Based upon the medium used to implement the wormhole attack, it is classified into the following two types:

   • *In-band*: In in-band wormhole attack, the same existing medium is used to establish the tunnel between two malicious nodes. In-band medium is the most common choice and is used by encapsulation, packet relay, and protocol deviation methods to launch the wormhole attack [6, 8].

- *Out-of-band*: An attacker in the out-of-band method does not use the same medium but different wireless network to achieve wormhole attack [8].

D. On the basis of the behavior of the sensor nodes while forwarding packets, visibility of intruders, and the identities of the wormhole nodes, the wormhole attack is categorized into three types: open, half-open, and closed wormhole attack.

- *Open wormhole*: The source and destination nodes as well as the two ends of the wormhole link are visible in open wormhole attack. The identities of the attackers are visible in the header of the data packets involved in route discovery. The sensor nodes think of the malicious nodes as direct neighbors although they are aware of the presence of these bogus nodes along the route.
- *Half-open wormhole*: One end of the wormhole link is visible, while the other is kept hidden. The contents of the data packets are not altered but are simply forwarded from end of the tunnel to the other end.
- *Closed wormhole*: The attacker in this mode creates a fake list of neighbors by creating an illusion that the distance between the source and the destination is of one hop. Also, the visibility of the source node, destination node, and two end nodes of the tunnel are not disclosed [2].

4 Techniques to Detect Wormhole Attack

Several techniques have been developed to detect the presence of wormhole attack. Some of these techniques are discussed below:

1. WRHT: The wormhole-resistant hybrid technique (WRHT) is a way of detecting wormholes in the sensor network using the concept of Delphi and watchdog methods. The hop count value and the wormhole presence probability are estimated by each of the source node in the network using this method. WRHT is an expansion to AODV protocol that tends to detect all categories of wormholes in the network with the help of information based on packet drop and packet delay for each hop and also for the complete route in the network [4].
2. Sinalgo Simulator: This technique uses the Sinalgo simulator to efficiently detect the presence of wormholes and increase network security as well as reliability while considering the characteristics and energy constraints of the wireless sensor network. To manage the sensor nodes, the method uses visiting center local (VCL) algorithm to divide the sensor network and form sectors iteratively. A mobile node has been introduced in order to differentiate between the legitimate and malicious nodes with the use of neighborhood discovery of

nodes. Upon receiving signals from source nodes, a sink directs the mobile agent to collect information from these nodes present in each sector and summarizes the collected and processed data and finally returns to the sink along with information being collected [1].

3. WGDD Algorithm: The wormhole geographic distributed detection (WGDD) algorithm is a mechanism to detect wormhole attack without using anchor nodes or specialized hardware. A procedure to count the number of hops of neighboring nodes is used. Once this hop counting process is implemented, a set of hop counts of the neighboring nodes that are one/k hop away is calculated for each of the sensor node. In order to obtain the shortest route, the sensor nodes run the Dijkstra's algorithm, and then with the help of multi-dimensional scaling, a local map is reconstructed at each of the node. Lastly, in order to detect wormholes in the network, the distortions in the local maps are identified with the help of a novel "diameter" feature [10].

4. Artificial Neural Network: In order to detect wormholes in both uniform and non-uniform environment, a mechanism has been proposed using artificial neural network. This technique is based upon neighborhood count and does not require any specialized hardware. In order to gather neighborhood counts, a node called detector node (which is mobile in nature) travels through random locations in the sensor network. When this detector node enters some location in the sensor network consisting of wormholes, there is a sudden increase in the number of neighbors present and this value is recorded in a data set. Thus, the overall data set contains the neighborhood count in the presence as well in the absence of wormholes in the network, which is in turn fed to the neural network for training and testing purposes. Once testing is over, the presence of wormholes in the sensor network is thus decided by the output of the artificial neural network [11].

5. Packet Leashes: A technique to detect and further defend the sensor network from wormhole attack is to use packet leashes. This mechanism imposes a maximum transmission range beyond which transmission of data packets is not allowed. Time delay and geographical location are the constant and independent physical parameters that are used to detect the presence of wormholes in the network. By using time synchronization as well as local information, the packet leashes exercises the maximum transmission range on forwarding of packets so as to prevent wormhole attack. The temporal leash is one type of packet leash where the lifetime of a packet is ensured by an upper bound. While forwarding data packets in a temporal leash, the sender appends the time at which the packet was sent. This value is then compared with the time at which the packet was received by the receiving node. Another type of leash is the geographical leash which imposes a limit on the distance between the sender and the receiver. The location of the sender as well as the time at which the packet was sent is included in the sent data packet. Computation of the upper bound on the distance between the sender and the receiver is carried out when the data packet reaches its destination [12].

6. MDS-VOW: The multi-dimensional scaling—visualization of wormhole (MDS-VOW) approach uses the concept of multi-dimensional scaling and visualization to detect wormholes present in the wireless sensor network. Firstly, the outline of the wireless sensor network comprising the sensor nodes is reconstructed using multi-dimensional scaling. The distance between each pair of sensor nodes is evaluated by providing the imprecise space between the sensor nodes that can hear each other as inputs, and thus, a virtual position for all the nodes is estimated. While estimating the distance, few errors occur in the reconstructed network whose adverse effects are minimized by using a surface smoothing technique. Finally, an analysis of the reconstructed network is carried out so as to detect the fake neighbors present in that network [13].

7. Directional Antennas: In order to mitigate wormhole attack, this technique prevents malicious nodes from creating false neighbors with the help of directional antennas. Since wormhole attack cannot be launched if the wormhole link is detected as fake and its requests are neglected, therefore a precise set of neighboring nodes is maintained for each of the sensor nodes. On the basis of the signals received, direction information can be collected by a sensor node using the directional antennas. Three protocols are designed in this approach that uses directional information to counter wormhole attack [14].

8. Transmission Range-Based Method (TRM): This is a real-world applicable technique used to efficiently detect wormhole attack without the need of any additional hardware. It uses the local neighborhood information to determine a network affected by wormhole attack, even in the presence of large transmission range. The TRM uses a directed graph with "N" sensor nodes to represent the network model. The bogus nodes and the normal nodes are categorized into two types and can be differentiated in terms of power, transmission range, and capability of computation. The network topology between two sensor nodes within a certain communication range is analyzed in order to detect wormholes between them with the help of geometric relationship of the nodes' location [15].

9. AOMDV Protocol: This technique uses ad hoc on-demand multipath distance vector (AOMDV) routing protocol to detect and defend the sensor network from wormhole attack. As soon as the routes between two nodes are decided, RREQ and RREP packets are used to evaluate the round-trip time (RTT) between them. For each of the route, the RTT is evaluated and then is divided by the respective hop count. Next, the threshold round-trip time is calculated by finding out the average RTT of all routes. A wormhole link exists in the sensor network if by comparing each RTT with the threshold value, the total RTT is less than the threshold RTT and that particular route has a hop count that is equal to two. The first neighboring node is considered as the wormhole node if the wormhole link does exist in that route and a dummy RREQ packet is forwarded through that route. The receiver detects the neighboring node as the wormhole node as it receives the dummy packet from it. As a result, the routing entries for these two nodes are removed from the source node and the same is broadcasted to all the other sensor nodes, thus blocking the wormhole link and making it useless [16].

10. Round-Trip Time: This method makes use of the transmission time and number of neighboring nodes of the sensor nodes in the wireless sensor network in order to identify wormholes present in the ad hoc on-demand routing (AODV) protocol. The detection procedure consists of three phases out of which the first step is to create a list of neighbors for each of the sensor node. The second step is to identify the path between the source node and the destination node. In the final phase, the position of the wormhole tunnel is located in order to take further action. The presence of wormholes is detected if there is an increase in the number of neighboring nodes within a radius along the value of RTT between successive nodes, while the length of the path between the nodes decreases [17].

11. Range-free localization: Using the concepts of range-free localization, two methods have been proposed in this paper in order to detect wormholes in the wireless sensor network. A scheme called "sensor localization with ring overlapping based on comparison of received signal strength indicator" (ROCRRSI) has been used as a range localization process for detection of wormholes in the network. The two methods that have been developed observe the irregularity in the network measurements at the physical layer and analyze the RSS value estimated by the nodes involved in localization process. The first strategy is implemented while the localization procedure is carried out by the sensor nodes, while the second strategy is implemented once the localization procedure has been done by the nodes [18].

12. Connectivity Information: This distributed technique detects the presence of wormhole attack in multi-hop wireless networks based on connectivity information. This is a localized algorithm wherein the wormholes are detected by locating forbidden structures in the connectivity graph of the network. The algorithm is developed with the help of a unit disk graph model and a general communication model where the communication model can be known or unknown. The basic idea is to locate structures in the graph that does not belong to a legitimate connectivity graph as embedding of the unit disk graph is not allowed by these illegal substructures. Although for all cases the algorithm will not be able to detect the wormhole attack, for connected networks, it guarantees a high detection accuracy [19].

13. Key Generation: An approach for detecting wormhole attack has been proposed in this paper by analyzing the number of packets that have been sent and received by the sensor nodes in the network. The method is divided into two phases where the first phase is to generate keys and the second one is to detect wormholes in the network. In the first phase, the generation of a secret key is carried out in order to prevent alteration of data by the illegitimate nodes. Once each of the sensor node locates their geographic location, they use a HELLO message to acquire

information regarding neighboring sensor nodes that are one hop away. The second phase emphasizes on gathering data from the sent and received packets of each of the sensor nodes. This gathered data is then validated so as to detect the presence of wormhole attack in the sensor network [20].

14. Neighborhood and Connectivity Information: In this approach, a protocol has been proposed in order to detect wormhole attack on the basis of neighborhood and connectivity information. With the use of a key management protocol that is secure, pre-distributive, and pair-wise in nature, this method is able to detect wormhole attack effectively for a minimal storage cost. There are three phases of the proposed method, out of which the first phase is to build the neighborhood table comprising of neighbors that are one hop away for each of the sensor nodes present in the network. The neighborhood table is extended to include the neighboring nodes that are two hops away during the second phase. The process of detecting the wormhole attack is carried out in the last phase. This detection procedure can be used with wireless sensor networks having constraints in resources [21].

15. Neighbor Discovery and Path Verification: This wormhole detection and prevention procedure are able to detect and also defend the presence of wormhole attack in the AODV protocol of the sensor network by discovery of neighboring nodes and verification of the paths taken by the sensor nodes. In order to make this algorithm work, a secure communication is established between the source and the destination nodes. The first phase of the approach is to discover routes of neighboring nodes where a HELLO message is shared among the neighboring nodes so that at each level, packets can be encrypted. In the second phase, a list of neighbors with one hop and two-hop transmission range is created so as to perform verification of the neighboring nodes in the network. The sensor network is said to be affected by the wormhole attack if there is an illegal entry of neighboring nodes that are two hops away [22].

16. Poster: This technique is developed on the basis of delay encountered during synchronized communication and thus emphasizes on the actual behavior of wireless sensor network to detect wormhole attack in duty cycling wireless sensor network. Considering the delay caused by the wormholes in the network, the proposed algorithm is implemented in two phases. In the first phase, detection is carried out through time synchronization by inspecting the time difference between reflective messages. During the second phase, a synchronized communication is used to detect the presence of wormhole attack on the basis of delay detected at the base station [23].

Table 1 presents brief summary of various techniques reviewed in the paper.

Table 1 Summary of techniques of wormhole attack detection

Technique	Description	Advantages	Disadvantages	Simulation results
Wormhole-resistant hybrid technique (WRHT) [4]	This is a hybrid technique to detect wormhole attack by calculating the probability of wormhole presence with the help of watchdog and Delphi schemes	Does not need additional hardware, detects wormhole attacks of all categories	Use of Delphi and watchdog schemes independently decreases performance of wormhole exposure	For 500 nodes: Accuracy analysis, acc = 0.98
Wormhole attack detection using Sinalgo simulator [1]	The method intends to detect wormholes in the network by introducing a mobile agent to differentiate between normal and malicious nodes	Enhances network lifetime by improving packet delivery rate and consumption of energy, maintains reliability of the communication process in the network	Mobility of sensor nodes is not considered, does not work for other sensor network attacks	During wormhole attack, for 200 nodes in 800 ms, increase in: energy consumption = 28% Packet drop ratio = 43%
Wormhole geographic distributed detection (WGDD) algorithm [10]	The algorithm detects wormhole attack by using a hop counting process. It then uses a multi-dimensional scaling for each node to construct local maps and finally a diameter feature to identify disruptions in network	Able to detect wormholes in irregularly shaped networks, low false detection and tolerance rates, estimates the approximate location of wormholes present	Detection rate for wormholes shorter than three hops decreases	For 2500 nodes: Overall detection rate = 100% (nearly) For shorter wormholes = 80%
Artificial neural network approach [11]	This technique detects wormhole attack in both uniform and non-uniform environments based on the concept of artificial neural network by counting the neighboring nodes	Leads to increase in detection rate and decrease in false positive rate, does not lead to communication overhead	Detection rate is better for uniform networks compared to non-uniform networks	For 500 nodes: Avg. detection accuracy = 98.25% False positive rate = 1.71% False negative rate = 0.02%

(continued)

Table 1 (continued)

Technique	Description	Advantages	Disadvantages	Simulation results
Packet leashes [12]	This procedure incorporates two types of packet leashes: temporal and geographical leashes along with a TIK protocol to detect wormhole attack	Computation overhead is low, use of TIK protocol results in instant authentication of received data packets, TIK along with a MAC layer protocol defend the network against replay, spoofing, and wormhole attack	Works only for single wireless transmission; for temporal leash, clocks should be tightly synchronized; for geographical leash, nodes must be aware of their location and clocks should be loosely synchronized	For geographical leash, reduced transmission range = 6.2 m
Multi-dimensional scaling—visualization of wormhole (MDS-VOW) [13]	The MDS-VOW mechanism uses multi-dimensional scaling for reconstruction of a layout with sensor nodes which is later subjected to visualization to detect wormholes	Ratio of false alarm regarding detection of wormholes is low; sensor nodes are not required to have specialized hardware; each of the steps involved can be enhanced independently	It is a centralized mechanism leading to a single point of failure, does not work in irregularly shaped network; sensor nodes are considered to be immobile	For 401 nodes, when error rate ≤ 0.6: Detection accuracy = nearly 100% Low false positive rate Low false negative rate
Directional antennas [14]	In this procedure, directional information is used by the sensor nodes to prevent the end points of a wormhole link from creating fake neighbors	Does not require local information or clock synchronization, minimizes packet collisions, less overhead on the network	Blocks connection of legitimate links, thus affecting network performance, does not defend against multiple end points wormhole attack	When density = 10, for verified neighbor discovery protocol: links lost < 0.5%, disconnected nodes = 0 For strict neighbor discovery protocol: links lost = 40%, disconnected nodes = 0.03%

(continued)

Table 1 (continued)

Technique	Description	Advantages	Disadvantages	Simulation results
Transmission range-based method (TRM) [15]	The transmission range of the packets is used as a medium to detect wormhole attack in this method based on the local information of neighboring nodes	Applicable in real-world instances as it does not require any special hardware, works even in large transmission range, and efficiency is not affected by dynamic topology	Applicable only for closed wormhole attack, and detection rate for other types of wormhole attack is not known	For 100 nodes: When side length of network is square, detection rate = 100% For spare network, detection rate is not 100%
Ad hoc on-demand multipath distance vector (AOMDV) protocol [16]	The technique uses AOMDV protocol to detect wormholes in the network with the help of round-trip time (RTT) mechanism	Specialized hardware is not required, this technique applies less overhead on the network, and the end to end delay is minimal	Assumes the sensor nodes to be stationary and does not consider dynamic environment, does not consider delays due to congestion and queuing	For 45 nodes: Avg. throughput = 286.4 bits/sec Avg. end to end delay = 80.8036 s Packet delivery fraction = 0.7892
Round-trip time [17]	The round-trip time and number of neighboring nodes are used in this technique to locate wormholes present in the AODV protocol	Minimal overhead on the network, less consumption of energy	Works only for homogenous, static, and symmetric networks	For 50 nodes: Detection rate = 100% when length of wormhole $\geq$ 5
Range-free localization [18]	This method employs two strategies based on range localization procedure to detect wormholes in the network	The detection process does not make use of any predefined parameters, does not require any additional hardware, and the algorithm is apt for wireless sensor network	Detection rate decreases in presence of shadowing effect.	For detection after localization, σ = 6 dB No. of anchors = 160 Probability of detection = 1 when probability of false alarm = 1

(continued)

Table 1 (continued)

Technique	Description	Advantages	Disadvantages	Simulation results
Connectivity information [19]	This method is able to detect wormhole attack in multi-hop wireless network with the help of connectivity information	The process is independent of the communication model used for wireless network, universally applicable as specialized hardware as well as localized information is not required	Does not guarantee detection of wormhole attack in all cases, and for cases with low density, the rate of detection decreases	For 144 nodes: When network disconnection probability = 0 and 50%, Detection rate = 100% (nearly) and 90%, respectively.
Key generation [20]	In this procedure, a secret key is generated first and then sent and received packets of each node are analyzed to detect wormholes present	Generation of secret key prevent packet alteration, and the dropping of packets even in the absence of wormhole attack is considered during simulation	The algorithm works only for out-of-band channel wormhole attack.	For two hops away nodes, wormhole attack detected when Max packet drop > ½ of sent packets
Neighborhood and connectivity information [21]	The information regarding the neighboring nodes and the connectivity of the sensor nodes is used as detection feature in this technique to identify wormholes	Additional hardware is not required, efficient in detecting wormhole attack with minimal cost in storage	The approach is not applicable for dynamic sensor networks	Detection accuracy = 95% Packet delivery ratio = 97.15% Throughput = 84kbps
Neighbor discovery and path verification [22]	This method tends to detect as well prevent the wormhole attack in the AODV protocol-based neighbor discovery and path verification	The method is implemented based on modified AODV protocol without any special hardware requirements	The packet delivery ratio can be further improvised	For 70 nodes: Throughput = 3.5 kbps Delay = 0.8 s Packet delivery = 80% (nearly)
Poster [23]	This method detects the presence of wormhole attack in duty cycling sensor network by observing delays in synchronized communication	Considers actual behavior of wireless sensor network	This method is able to detect only wormholes launched by pair of hidden and collaborative sensor nodes	Evaluation remains for future work

5   Conclusion

Wireless sensor networks are prone to several kinds of attacks. Wormhole attack is a serious one, as it distorts routing information and further disables the network, even for an authenticated and encrypted one. This paper has critically reviewed the wormhole attack vis-à-vis its working, effects, and types along with several techniques proposed in the literature to detect wormhole attack in WSN.

Although considerable work has been done in developing mechanisms for detection and mitigation of wormhole attack, the parametric evaluation of the performance of these techniques calls for further research.

References

1. Bendjima, M., Feham, M.: Wormhole attack detection in wireless sensor networks. In: Proceedings of SAI Computing Conference, London, UK, 13–15 July 2016
2. Maidamwar, P., Chavhan, N.: A survey on security issues to detect wormhole attack in wireless sensor network. Int. J. AdHoc Netw. Syst. 2(4), 37–50 (2012)
3. Pawar, R.B., Patil, P.U., Bombale, G., Zalani, A.: Wormhole attack and it's variants in wireless sensor network: a survey. Int. J. Eng. Res. Technol. 3(8), 1176–1179 (2014)
4. Singh, R., Singh, J., Singh, R.: WRHT: a hybrid technique for detection of wormhole attack in wireless sensor networks. Mobile Inf. Syst. 8354930, 13 (2016)
5. Poonam, M.: Wormhole attack in wireless sensor network: a survey. Int. J. Adv. Res. Sci. Eng. 5(2), 110–117 (2016)
6. Gupta, A., Gupta, A.K.: A survey: detection and prevention of wormhole attack in wireless sensor networks. Glob. J. Comput. Sci. Technol.: E Netw. Web Secur. 14(1), 23–31 (2014)
7. Ladva, M.M., Lathigara, A.M.: Wormhole attack detection and prevention technique in mobile ad-hoc network: a review. Int. J. Innov. Emerg. Res. Eng. 2(2), 83–88 (2015)
8. Ughade, S., Kapoor, R.K., Pandey, A.: An overview on wormhole attack in wireless sensor network: challenges, impact and detection approach. Int. J. Recent Develop. Eng. Technol. 2 (4), 105–110 (2014)
9. Sharma, N., Singh, U.: Various approaches to detect wormhole attack in wireless sensor network. Int. J. Comput. Sci. Mobile Comput. 3(2), 29–33 (2014)
10. Xu, Y.,Chen, G., Ford, J., Makedon, F.: Detecting wormhole attacks in wireless sensor networks. In: Goetz, E., Shenoi, S. (eds) Critical Infrastructure Protection. International Federation for Information Processing, pp. 267–279. Springer Series in Computer Science, Springer, Berlin (2008)
11. Shaon, M.N.A., Ferens, K.: Wireless sensor network wormhole detection using an artificial neural network. In: International Conference of Wireless Networks, pp. 115–120, Las Vegas, USA (2015)
12. Hu, Y.C., Perrig, A., Johnson, D.B.: Wormhole attacks in wireless networks. IEEE J. Sel. Areas Commun. 24(2), 370–380 (2006)
13. Wang, W., Bhargava, B.: Visualization of wormholes in sensor networks. In: Proceedings of 3rd ACM workshop on wireless security (WiSe'04), pp. 51–60. Philadelphia, USA (2004)
14. Hu, L., Evans, D.: Using directional antennas to prevent wormhole attacks. In: Proceedings of Network and Distributed System Security Symposium (NDSS 2004), San Diego, California, USA (2004)
15. Wu, G., Chen, X., Yao, L., Lee, Y., Yim, K.: An efficient wormhole attack detection method in wireless sensor networks. Comput. Sci. Inf. Syst. 11(3), 1127–1141 (2014)

16. Parmar, A., Vaghela, V.B.: Detection and prevention of wormhole attack in wireless sensor network using AOMDV protocol. In: Proceedings of 7th International Conference on Communication, Computing and Virtualization, vol. 79, pp. 700–707. Mumbai, India (2016)
17. Tun, Z., Maw, A.H.: Wormhole attack detection in wireless sensor networks. World Acad. Sci. Eng. Technol. **46**, 545–550 (2008)
18. García-Otero, M., Población-Hernández, A.: Detection of wormhole attacks in wireless sensor networks using range-free localization. In: Proceedings of 17th International Workshop on Computer Aided Modelling and Design of Communication Links and Networks (CAMAD), pp. 21–25, Barcelona, Spain (2012)
19. Maheshwari, R., Gao, J., Das, S.R.: Detecting wormhole attacks in wireless networks using connectivity information. In: Proceedings of 26th IEEE International Conference on Computer Communications, Barcelona, Spain (2007)
20. Buch, D., Jinwala, D.: Detection of wormhole attacks in wireless sensor network. In: Proceedings of International Conference on Advances in Recent Technologies in Communication and Computing, pp. 7–14, Bangalore, India (2011)
21. Patel, M., Aggarwal, A.: Detection of hidden wormhole attack in wireless sensor networks using neighbourhood and connectivity information. Int. J. Ad hoc Netw. Syst. **6**(1) (2016)
22. Johnson, M.O., Siddiqui, A., Karami, A.: A wormhole attack detection and prevention technique in wireless sensor network. Int. J. Comput. Appl. **174**(4), 1–8 (2017)
23. Minohara, T., Nishiyama, K.: Poster: detection of Wormhole attack on wireless sensor networks in duty-cycling operation. In: Proceedings of the 2016 International Conference on Embedded Wireless Systems and Networks, pp. 281–282. Graz, Austria (2016)

A General Framework for Spectrum Assignment in Cognitive Radio Networks

Monisha Devi, Nityananda Sarma and Sanjib Kumar Deka

Abstract With its opportunistic nature to exploit the unused licensed bands, cognitive radio (CR) presents itself as an advancing technology that senses for spectrum holes and distributes them rightly among secondary users. This paper proposes a spectrum assignment approach for CR networks that looks to provide a proper sharing of the available channels such that spectrum utilization is improved. While incorporating a concurrent strategy for channel allocation, the model constraints to the allocation limitation stating that one channel can be allotted to only one user at a time, and also one secondary user can gain access to at most one channel at a time. Further, simulation results of the proposed model show its effectiveness in terms of spectrum utilization efficiency.

Keywords Cognitive radio · Dynamic spectrum access · Secondary user
Primary user · Spectrum opportunity

1 Introduction

With the proliferation of wireless technology, the problem of radio spectrum shortage where applications run out of frequencies becomes universal. The static spectrum assignment policy of government bodies' permits allocation of spectrum bands to licensed users such that no other user gains access during the licensed duration. However, spectrum utility measurements captured by Federal Communications Commission (FCC) report that a significant portion of licensed

M. Devi (✉) · N. Sarma · S. K. Deka
Department of Computer Science & Engineering, Tezpur University,
Tezpur 784028, India
e-mail: mnshdevi@gmail.com

N. Sarma
e-mail: nitya@tezu.ernet.in

S. K. Deka
e-mail: sdeka@tezu.ernet.in

© Springer Nature Singapore Pte Ltd. 2019
J. K. Mandal et al. (eds.), *Advanced Computing and Communication Technologies*,
Advances in Intelligent Systems and Computing 702,
https://doi.org/10.1007/978-981-13-0680-8_15

spectrum remains unutilized by the licensed owners. Such ineffectual use of spectrum resource brings about an innovative technology called cognitive radio (CR) [2] leading to the concept of dynamic spectrum access. CR features a dynamic behaviour wherein temporarily vacant spectrum bands are made available opportunistically to the secondary users (SUs) while restraining the primary users (PUs) from any harmful interference. With the intent of enhancing spectrum utilization, CR includes spectrum sensing, spectrum management, spectrum sharing and spectrum mobility as its key functionalities. The spectrum holes obtained during sensing are taken as input for the sharing phase where allocation mechanisms plan to provide a fair distribution of the idle channels among SUs. As such, spectrum sharing [2] looks for proper design models that can make the utmost use of idle bands by proffering efficient channel assignments.

In this paper, a spectrum assignment mechanism is formulated for CR networks (CRN) that allows a central entity to take over the control of channel allocations. The approach takes into consideration the dynamics in spectrum opportunity (SOP) so that the same set of channels is not available to every SU. Also, the approach follows a concurrent strategy where all vacant channels are to be shared at the same time while handling the allocation limitation that checks that one channel can be allocated to only one SU at a time, and also one SU can exploit at most one channel at a time. For performance evaluation, simulations are carried out in terms of spectrum utilization and throughput, and the results are compared with a random allocation strategy.

The rest of the paper is organized as follows. Section 2 covers related works. Sections 3 and 4 discuss the proposed model and performance evaluation, respectively. And finally in Sect. 5, conclusions are drawn based on simulation and pointers are provided for future work.

2 Related Work

Spectrum sharing plays a vital role in CR networks that allows the vacant channels to be fairly allocated while keeping up with some prerequisite conditions in the designed model. A study on channel assignment models is carried out in [1]. Deploying a bipartite graph for spectrum allocation is discussed in [5] where winner determination for the available channels is designed using maximal weight bipartite matching in the graph. Hyder et al. [4] formulated two online auction models for allocating the spectrum among unknown number of bidders (SUs). A game theoretical model is designed in [7] where solution to the problem of spectrum allocation is defined using the Nash equilibrium. The concept of evolution is taken up for planning the allocation in a different direction [8] where arrangement of a chromosome string represents a conflict-free channel allocation. Dong et al. [3] discussed a double-sided spectrum auction where both seller-side and buyer-side carry out their winner determination in a dissembled way and then winner sets are merged to provide the allocation status. A multiple knapsack problem formulation for channel allocation is studied in [6] that looks for maximizing the spectrum utilization. With a state of the art on spectrum sharing approaches, it can be noticed that most of the work does not take into account one of the prime concern of CR networks, that is, the dynamics in channel availability among SUs as to which the same set of channels may not be available to every SU. Also, this work enables a central entity to determine the channel allocation decisions. With a sequential strategy where channels are shared one by one, a SU who is allotted a channel may have a higher willingness for another vacant channel. But since the SU has already won a channel, it cannot participate further in the sharing process. So, the designed model allows a concurrent strategy where all idle channels are shared away simultaneously so that every SU can show its willingness for every channel while obeying to the channel availability constraint.

3 Spectrum Assignment in CRN

3.1 System Model

The designed model deploys a cognitive radio network, consisting of SUs, which coexist with a primary network, consisting of PUs. The PUs are licensed users who want to give up their unused channels so that SUs can benefit from them. It is assumed that M number of channels are available among N number of SUs such that $N > M$. The approach allows a central entity, C_e, to take over the channel assignment decisions such that it meets the allocation and availability constraints. With the allocation constraint, one channel can be assigned to only one SU at a time, and also one SU can exploit at most one channel at a time. And with the availability constraint, a SU can gain access to a channel only when it is sensed as available by the SU. All SUs are considered to be within the transmission range of C_e. Channels are sensed by the SUs before starting the allocation process since channel availability of a SU varies with respect to PUs' locations. Then accordingly, a SU i maintains a list $C_i = \{c_i^1, c_i^2, ..., c_i^j, ..., c_i^M\}$ such that $c_i^j = 1$ if channel j is sensed as available by SU i, otherwise $c_i^j = 0$. A channel availability matrix, X, controls the SOP lists of all SUs at C_e such that $X = \{x_{ij} | x_{ij} \in \{0, 1\}\}_{N \times M}$ where $x_{ij} = 1$ if $c_i^j = 1$ in C_i, otherwise when $c_i^j = 0$, $x_{ij} = 0$. Again, a channel allocation matrix, A, represents the channel assignments obtained during the process where $A = \{a_{ij} | a_{ij} \in \{0, 1\}\}_{N \times M}$, $a_{ij} = 1$ if SU i gets to use channel j and $a_{ij} = 0$ otherwise. Moreover, C_e manages the winner SUs for the available channels in a list $W = \{w_j\}_{1 \times M}$ where $w_j = i$ if $a_{ij} = 1$ implying that winner SU i gets to exploit the channel j. Otherwise, $w_j = 0$ meaning channel j is unallocated. Initially, since all channels are free, winners are set to 0. For exchanging control messages, a common control channel (CCC) is considered in the network.

3.2 Proposed Scheme

With the available channels being shared out concurrently, it enables every SU to unveil its willingness for the channels according to their channel availability constraint. The central entity C_e learns about the channels, $C_{List} = \{1, 2, 3, ..., M\}$, that are left free by the PUs, and looks to allocate them among SUs. Before starting the allocation, a SU i senses for the vacant channels and manages them in C_i. Thereafter, in keeping with the SOP list, SU i decides upon the reward values for every available channel in C_i. It is considered that the reward values can represent the bandwidth that SU i can acquire by using the channels. As such for SU i, $B_i = \{b_i^1, b_i^2, ..., b_i^j ....., b_i^M\}$ builds up where b_i^j represents the reward value of SU i for channel j. However, if j is unavailable at SU i, i.e. $c_i^j = 0$, then $b_i^j = 0$. To organize the reward values from SUs at C_e, a reward matrix, R, is arranged such that $R = \{r_{ij}\}_{N \times M}$ where

$$r_{ij} = \left\{ \begin{array}{ll} b_i^j & \text{if } c_i^j = 1, \text{ where } c_i^j \in C_i, i \in \{1, 2, ..., N\}, j \in \{1, 2, ..., M\} \\ 0 & \text{otherwise} \end{array} \right\}$$

(1)

Now, every SU i informs about its SOP list C_i and reward list B_i to the C_e and C_e accordingly arranges them in channel availability matrix, X, and reward matrix, R respectively. As such C_e gathers the knowledge of SOP and reward values of all SUs. The degree of a SU i is taken to be the number of unassigned channels in its C_i which can be given as follows:

$$\text{Degree}(i) = \sum_{\substack{j = 1 \\ w_j = 0}}^{M} c_i^j \ \forall \ i \in \{1, 2, ..., N\}$$

(2)

Also, two different states, winner and competitor, are assumed for the SUs of the network and every SU will be in either one of the two states. Initially, all SUs are in the competitor state when they are not being assigned any channel. C_e computes the degree of every competitor SU after assigning a channel to a winner SU. To start with the channel assignment strategy, a SU i who is in need of a channel is randomly picked and it proceeds for allocation as described in Algorithm 1. After a channel gets allotted to a SU, it is checked whether the C_{List} is empty implying that there are no more free channels. If the condition gets to be true, it suggests that the channel assignment process completed such that all idle channels are assigned rightly among the users. But if C_{List} is not empty, then degree of every competitor SU is checked. If at least one competitor SU has degree greater than 0, then the channel assignment strategy proceeds. Otherwise, in very few cases it may appear that the degree of every competitor SU happens to be 0 when C_{List} is non-empty. Such case arises when competitor SUs do not have the unallocated channels in their

SOP lists due to which even if a channel is free it cannot be allotted. In such situation, take every unused channel j and do the following. The jth column of matrix X is looked so that if $\sum_{i=1}^{N} x_{ij} = 0$, it implies that channel j is not available in SOP list of any SU. So, the channel shall remain as unallocated. Or else, get the SU k that has highest reward value for j. Surely k has to be a winner SU. Now, the channel assigned to SU k is obtained, say m, and all the SUs that have submitted reward values for channel m, except k, are looked for. If there exists at least one competitor SU among them, then the competitor SU with higher value is considered and channel m goes to the competitor SU and j is given to SU k. Otherwise, if all SUs are in winner state, then value next to the winning value r_{km} is taken. If l is the SU whose value is next to r_{km}, the channel assigned to SU l is picked and same process gets repeated. On finding a competitor SU during the process, SU and channel pairs are swapped such that SU k gets the channel j. But, if during the process it is found that for a channel the reward value submitted is only from the winner SU, implying that no other SU has that channel in their SOP list, then in such case channel j cannot be allotted to SU k and so the SU having the next highest reward value for channel j is selected. Incorporating such method may somewhat reduce the spectrum utility since a channel in such case may not go to the highest valued SU. But, it tries to make utmost use of the available channels at the end while adhering to the allocation constraint.

Algorithm 1: Channel Assignment Algorithm

Input: Reward matrix R having reward values for all channels from every SU
Output: Channel assignments given in allocation matrix A and winner list W
Steps:
1: *while* (notEmpty(C_{List}) $\vee$ (Degree(competitor SU) $\neq$ 0)), *do*
2:     Randomly pick a SU.
3:     Look into the status of SU.
4:     *If* SU is a winner, pick another SU and goto Step 3. *Else* goto Step 5.
5:     Get the channel for which the chosen SU has highest value.
6:     Look into the winner list W.
7:     *If* channel is already allocated, select channel with next highest value and goto
        Step 6. *Else* goto Step 8.
8:     Look into the reward matrix R.
9:     *If* reward value for the channel from the chosen SU is greater than equal to reward values
        from all other SUs, goto Step 10. *Else* goto Step 11.
10:    Channel is assigned to the chosen SU, and allocation matrix A is updated. Assigned
        channel is removed from C_{List}. Goto Step 1.
11:    Get the SU that has a higher value for the channel than the chosen value and look into
        its status.
12:    *If* SU is a winner, goto Step 13. *Else* goto Step 14.
13:    *If* there exists any other SU with value for the channel greater than the chosen value,
        goto Step 11. *Else* goto Step 10.
14:    Channel is left unallocated. Goto Step 2.
15:    *end*
end

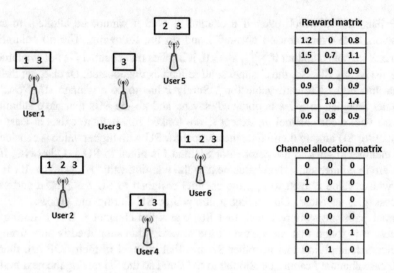

Fig. 1 An example for the proposed spectrum assignment approach

An example of the designed model is shown in Fig. 1 that takes a CR network with six SUs and three vacant channels. Reward values are provided in the reward matrix, and allocation statuses are in the channel allocation matrix. SU $S2$ is randomly picked whose highest rewarded channel is $C1$. Since r_{21} is the highest among other values for $C1$, channel $C1$ assigned to $S2$. Next, $S6$ is taken who gave the highest value to channel $C3$. But for $C3$, r_{53} is the highest reward and SU $S5$ is still a competitor. So, $C3$ is not allocated to $S6$ and another SU is looked. $S4$ takes $C1$ into consideration (randomly picked between $C1$ and $C3$ since same reward value), but as $C1$ is already allotted, it takes $C3$. However, again there is no allocation for $C3$. Then, on getting the SU $S5$, channel $C3$ gets assigned to it. Now when $S6$ is picked, $C3$ is already allocated to its highest rewarded SU, and $S6$ gets the next highest rewarded channel $C2$. Although $C2$ has highest reward from $S5$ but since it is in winner state, $S6$ gets to use $C2$.

4 Performance Evaluation

To evaluate the performance of the proposed spectrum sharing approach, two different network scenarios are explored where, in one scenario, number of channels are kept fixed at 6 and number of SUs are varied as $\{20, 30, 40, 50, 60\}$, and in the other scenario, number of SUs are kept fixed at 40, and number of channels are varied as $\{4, 6, 8, 10, 12, 14\}$. Simulations are executed using MATLAB R2013a in Windows® environment that deploys the network in an area of 600 m × 600 m and averages the results over 100 rounds. A comparison of the designed model is made with a random allocation strategy so as to assess the performance improvement in terms of spectrum utilization and throughput. With a random allocation, a SU who randomly selects a channel is given to exploit the channel if availability and allocation constraints are met. Considering spectrum utilization and throughput as the performance metrics, spectrum utilization, S_u, is defined as the sum of allocated reward values for the available channels, i.e. $S_u = \sum_{j=1}^{M} \sum_{i=1}^{N} a_{ij} r_{ij}$.

For computing the throughput, T_r, acquired by the allocation, it is assumed that all channels exhibit the same transmission power, $P_T = 0.01$. And accordingly, $T_r = r_{ij} \log_2 \left(1 + P_T \left(\frac{PL_i}{I_i + \sigma^2}\right)\right)$. σ^2 is the noise variance and taken as 10^{-5} for all, PL_i is the path loss factor between SU i's transmitter and receiver, I_i is the interference from PUs. r_{ij} is the reward value of winner SU i for channel j.

With Fig. 2, a comparison of spectrum utilization, S_u, is shown when number of channels are varied for different sets of SUs. Here it can be observed that with increasing number of channels, S_u also increases for all SU sets. This is because when the count of channels gets larger, it implies that more spectrum resource is available for the users to operate upon. And as such, more number of SUs can get access to the spectrum. However, with changing number of channels, S_u for different SU sets remains more or less the same. Although with increasing number of SUs, a wider range of reward values may appear which may cause a slight growth in S_u, but mostly the spectrum utility remains roughly similar for different SU sets.

In Fig. 3a, a comparison of the proposed model in terms of spectrum utilization, S_u, is considered with the random allocation when the number of SUs is kept to be varying. S_u gives a reasonable performance for the proposed model since it tries to allocate the channels to the SUs with higher reward values. But in situations where some free channels are left out even when SUs are unassigned, channel allocation to

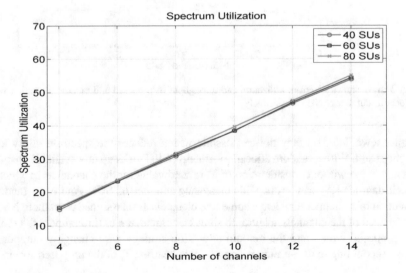

Fig. 2 Spectrum utilization for different sets of SUs with varying number of channels

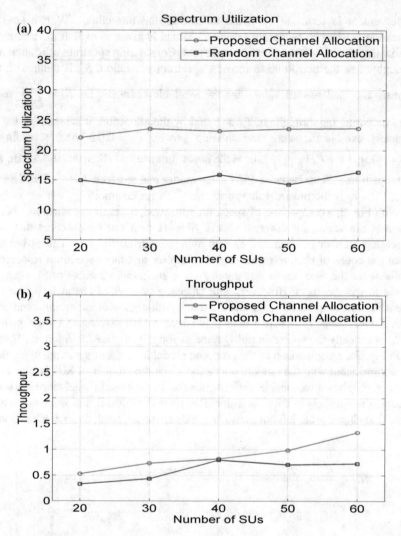

Fig. 3 a, b Shows spectrum utilization and throughput of proposed and random allocation with respect to number of SUs, respectively

higher rewarded SUs may not be possible which reduces the spectrum utility to a slight extent. With a random allocation strategy, a SU picks up any available channel for allocation without consideration of its reward value for the channel as to which a much degraded S_u can result with increasing number of SUs. And also, random selection of a channel may leave some idle channels as unassigned even though SUs are in need of the channels. Figure 3b shows a comparison of throughput, T_r, among the proposed model and the random allocation strategy with changing number of SUs. Accounting to the similar reasons as stated earlier, T_r of the proposed spectrum sharing approach displays a finer performance compared to random channel allocation wherein SUs may randomly prefer to go for less rewarded channels and even keep away the channels as unassigned due to unsuitable allocation strategy.

Again in Fig. 4a, spectrum utilization, S_u, is compared between designed model and random allocation when number of channels is made to vary. A better performance is shown by the proposed mechanism since channels being chosen randomly by a SU may not be of a higher value. However, in most cases, S_u shows a growth for both the strategies since more channels become available for the SUs.

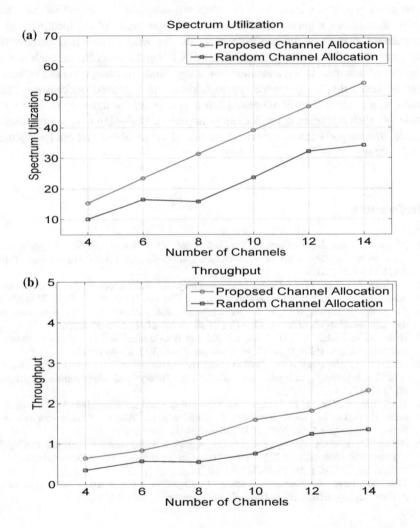

Fig. 4 **a, b** Shows spectrum utilization and throughput of proposed and random allocation with respect to number of channels, respectively

Similar situation occurs with the throughput, T_r, as can be seen in Fig. 4b, where the proposed approach shows an acceptable performance over random allocation while changing the number of channels. The model planned in this approach tries to make the utmost use of the available channels while considering the design constraints.

5 Conclusion

The paper proposes a framework for spectrum assignment in cognitive radio network that allows a proper distribution of the spectrum. While incorporating a concurrent sharing of the available channels, the model takes into account the dynamic channel availability characteristic of CRN and awards the channels among rightful SUs subject to the allocation limitation. From the design and evaluation, it can be stated that the proposed approach shows an improved performance compared to a random channel allocation and tries to make the most use of the vacant channels such that spectrum utility can be improved. Designing of spectrum sharing model that supports channel reuse can be one of the problems that can be explored in the future.

References

1. Ahmed, E., Gani, A., Abolfazli, S., Yao, L.J., Khan, S.U.: Channel assignment algorithms in cognitive radio networks: taxonomy, open issues and challenges. IEEE Commun. Surv. Tutor. **18**(1), 795–820 (2016)
2. Akyildiz, I.F., Lee, W.Y., Vuran, M.C., Mohanty, S.: NeXt generation/dynamic spectrum access/cognitive radio wireless networks: a survey. Comput. Netw. **50**(13), 2127–2159 (2006)
3. Dong, W., Rallapalli, S., Qiu, L., Ramakrishnan, K.K., Zhang, Y.: Double auctions for dynamic spectrum allocation. IEEE/ACM Trans. Netw. **24**, 2485–2497 (2016)
4. Hyder, C.S., Jeitschko, T.D., Xiao, L.: Bid and time truthful online auctions in dynamic spectrum markets. IEEE Trans. Cogn. Commun. Netw. **3**(1), 82–96 (2017)
5. Khaledi, M., Abouzeid, A.A.: Auction-based spectrum sharing in cognitive radio networks with heterogeneous channels. In: Information Theory and Applications Workshop (ITA) (2013)
6. Li, C., Liu, W., Liu, Q., Li, C.: Spectrum aggregation based spectrum allocation for cognitive radio networks. In: IEEE International Conference on Wireless Communications and Networking (WCNC), pp. 1626–1631. IEEE (2014)
7. Niyato, D., Hossain, E.: A game-theoretic approach to competitive spectrum sharing in cognitive radio networks. In: IEEE International Conference on Wireless Communications and Networking (WCNC), pp. 16–20. IEEE (2007)
8. Zhao, Z., Peng, Z., Zheng, S., Shang, J.: Cognitive radio spectrum allocation using evolutionary algorithms. IEEE Trans. Wirel. Commun. **8**, 4421–4425 (2009)

Novel Schemes for Energy-Efficient IoT

Kandikonda Venkateshwarlu and Pushparaj Shetty D

Abstract Internet of things (IoT) is a global infrastructure for the information society which enables advanced services by interconnecting physical and virtual things based on existing and evolving inter-operable information and communication technologies. Developing green IoT is a difficult task because IoT has more devices and has complex structure, so most of the current schemes for deploying nodes in wireless sensor networks (WSNs) cannot be applied directly in IoT. In this paper, we propose a scheme which gives an energy-efficient IoT. Here, we propose two schemes for framework structure of a network, and then we propose clustering algorithms and routing algorithms for network formation which is based on minimum spanning tree. After numerous simulations, we show that these schemes result in minimal energy consumption and enhance the network lifetime. Thus, the proposed schemes are more energy-efficient compared to a typical WSN deployment scheme; hence, these schemes are applicable to the green IoT deployment. We show that in the proposed schemes, the nodes are alive for more number of rounds as compared to the existing algorithms.

Keywords Internet of things · Wireless sensor networks · Energy efficiency Clustering · Minimum spanning tree · Steiner tree

1 Introduction

The Internet of things (IoT) is the inter-networking of smart objects. The expected properties of an IoT are offering advanced connectivity of devices, systems, and services. IoT has applications in several domains such as health care, smart city.

K. Venkateshwarlu · P. Shetty D (✉)
Department of Mathematical and Computational Sciences,
National Institute of Technology Karnataka, Surathkal 575025, India
e-mail: prajshetty@nitk.edu.in

K. Venkateshwarlu
e-mail: kvenkateshwarlu.2010@gmail.com

© Springer Nature Singapore Pte Ltd. 2019   173
J. K. Mandal et al. (eds.), *Advanced Computing and Communication Technologies*,
Advances in Intelligent Systems and Computing 702,
https://doi.org/10.1007/978-981-13-0680-8_16

It is expected that there will be a steep rise in the automation of these embedded smart devices in all fields. These devices have to be identified individually by assigning an IP address and should be able to transmit and receive data over a network without manual intervention. The embedded technology in the objects helps them to interact with internal states or the external environment, which in turn affects the decisions taken. IoT should support remote accessing and integration of physical- and computer-based systems. As there is a continuous data transmission between the devices, there is a need for energy-efficient communication among the sensor nodes. Further, as the number objects deployed in IoT are more, a large amount of power is consumed, so green networking plays a crucial role in IoT to reduce the power consumption. Because of continuous sensing property, sensors consume more energy. An important objective in the design of IoT is achieving energy efficiency, as the power sources for sensors are of limited capacity.

We use a clustering algorithm which uses the *k means* clustering to address the problem of optimizing communication cost. We show that the proposed scheme can work more efficiently compared to other WSN deployment schemes; thus, it is applicable to green IoT deployment. In the first scheme, the area is divided into groups which are called clusters, and communication between nodes can be done within the cluster only. Among clusters, communication is possible through relay nodes. We use clustering and routing algorithm for building the network. Then, we perform a number of experiments with varying number of nodes to test the scalability of the proposed scheme. The results show that the proposed scheme is more energy-efficient as compared to existing ones.

2 Related Works

Many clustering-based routing protocols are proposed for IoT applications. Hybrid Energy-Efficient Distributed (HEED) clustering [1] periodically selects cluster heads according to a combination of two or several parameters like proximity or node degree. Power-Efficient Gathering in Sensor Information System (PEGASIS) protocol is a chain-based routing scheme. Low-Energy Adaptive Clustering Hierarchy (LEACH) is a TDMA-based MAC protocol which integrates clustering and a simple routing protocol in WSNs. Hierarchical Cluster-based Routing (HCR) technique is an extension of the LEACH protocol that is a set of organized cluster-based approach for continuous monitoring [2]. Stable Election Protocol (SEP) is based on weighted election probability of each node to become cluster head according to the remaining energy in each node. EEHCA an Energy-Efficient Hierarchical Clustering Algorithm for WSNs achieves a good performance in terms of lifetime by minimizing energy consumption for communication. EECS [3] is Energy-Efficient Clustering Scheme for WSNs, DEEC [4] is Distributed Energy-Efficient Clustering protocol, Modified Low-Energy Adaptive Clustering Hierarchy (MODLEACH), cross-layer protocol are few of the available clustering algorithms. Other clustering energy efficient protocols for WSN are elaborated in references [5–10]. There are two important energy-efficient models proposed in the literature for IoT. First one is a framework for green IoT proposed by Huang et al. [11]; the other framework is proposed by Rani et al. [12].

3 System Framework

3.1 First Scheme

In the first scheme [11], the *k means* clustering algorithm is used. The nodes are deployed in the selected area. Using *k means* clustering algorithm, the nodes are divided into clusters, and each of these clusters is numbered with ids. These ids will help in forming connections easily. Each cluster will have several nodes. Each node is given an initial energy which will be used for various tasks like clustering formation, routing, data transmission, and reception. Each node has communicating range r, which means it can only communicate with nodes which are within that range and which have the same cluster id. Then, the relay nodes are deployed at the center of the cluster. In this algorithm, nodes are allowed to communicate with the relay nodes only. This will decrease unnecessary data transmissions and will result in energy efficiency. Similarly, relay nodes also have initial energy, which is more as compared to that of normal nodes. Relay nodes have communication range R, so they will only communicate with other relay nodes, which are in that range. Relay nodes which are closer to the base stations are connected directly.

After this, we find the Steiner points of the network. Steiner tree problem is an algorithmic problem of finding extra "Steiner" points to be added to a point set in order to reduce the cost of connecting the points. Most versions of the Steiner tree problem are NP-hard, but some restricted cases can be solved in polynomial time. The theory of Steiner tree problems is found in [13]. The Steiner tree problem in graphs can be treated as a generalization of two problems, namely the shortest path problem and the minimum spanning tree (MST) problem (Fig. 1).

Consider two nodes x and y with $p(x, y)$ as the distance between them. In each cluster, normal nodes form group N, relay nodes form group RN, base stations form group BS, and the entire cluster heads form group CH. Id of each cluster is denoted by Cid. Whole network is denoted by $H(V, W)$, where V is set of all the nodes in the network and W is the set of links between the nodes. Relay nodes can only communicate with other relay nodes or which are in its communication range R, similarly normal nodes can only communicate with relay nodes which are within its

Fig. 1 **a** Steiner tree of three points A, B, and C. S is a Steiner point, **b** Steiner tree of four points A, B, C, and D. S1 and S2 are Steiner points

communication range, say, r. Here, relay node's communication range $R \geq$ normal node's communication range. Sending and receiving the data between the nodes and relay nodes can be done only when the following conditions are satisfied:

For intra-cluster communication, data transmission and reception occur if: $x \in N, y \in RN, p(x, y) \leq r$: $x, y \in RN, p(x, y) \leq r$ and does not occur if: $x, y \in N, p(x, y) > r$: $x \in N, y \in RN, p(x, y) > r$: $x, y \in RN, p(x, y) > R$. Nodes transmit data to the base station through relay nodes by following multi-hop communication.

3.2 Second Scheme

In the second scheme, the algorithm used for the clustering and routing is layered clustering algorithm. The deployment and clustering are done as in scheme 1. Each cluster will have several numbers of nodes and relay nodes. In this algorithm, nodes are allowed to communicate with the relay nodes only. Connection between relay nodes is done using routing algorithms. Relay nodes which are closer to the cluster heads are connected directly. After this, cluster heads will join cluster coordinators from the immediate adjacent layer above and so on. An illustration of this scheme is given in Fig. 2.

In each cluster, the coordinators between clusters form the group CCO, all the cluster heads form group, say, CH. A cluster which is above is named as T_CL, and cluster which is below is named as B_CL.

Fig. 2 Layered structure of cluster network

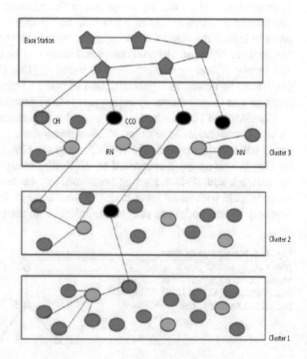

For intra-cluster communication, data transmission and reception will occur if: $x \in N, y \in RN, p(x,y) \leq r$: $x \in CH, y \in RN, p(x,y) \leq r$ and will not occur if: $x, y \in N$, $p(x,y) > r$: $x \in N, y \in CH, p(x,y) > r$: $x, y \in RN, p(x,y) > R$. Multi-hop communication is used to transmit data to the base station through relay nodes. For inter-cluster communication, data transmission and reception will occur if:

$x \in CH$ from B_CL, $y \in CCO$ from $T\_CL, p(x,y) \leq r$,

$Cid(B\_CL) = Cid + 1(B\_CL)$

$x \in CH$ from B_CL, $y \in CH$ from $T\_CL, p(x,y) \leq r$,

$Cid(B\_CL) = Cid + 1(T\_CL)$

$x \in CH$ from $B\_CL, y \in BS$ from $T\_CL, p(x,y) \leq r$,

$Cid(B\_CL) = Cid + 1(T\_CL)$ and will not take place if:

$x \in N$ from $B_{CL}, y \in N$ from $T_{CL}, p(x,y) > r$:

$x \in N$ from $B_{CL}, y \in RN$ from $T_{CL}, p(x,y) > r$:

$x \in RN$ from $B_{CL}, y \in RN$ from $T_{CL}, p(x,y) > R$:

$x \in N$ from $B_{CL}, y \in CH$ from $T_{CL}, p(x,y) > r$:

$x \in CH$ from $B_{CL}, y \in CH$ from $T_{CL}, p(x,y) > r$:

$x \in RN$ from $B_{CL}, y \in CH$ from $T_{CL}, p(x,y) > r$:

$x \in CH$ from $B_{CL}, y \in CH$ from $T_{CL}, p(x,y) > r$.

4   Modeling the Network

4.1   First Scheme

In this framework, the structure of the network is explained in detail. In the network, all the nodes are divided into different groups which are called clusters, to reduce the data transmission and reception distance between the nodes and thereby reducing the power consumption. This algorithm reduces the power consumption by using *k means* clustering.

Algorithm 1

Step 1: First deploy the nodes (N) in a fixed area.

Step 2: Apply *k means* algorithm on the nodes for cluster formation.

Step 3: Deploy the relay nodes at the center of the clusters. And fix the base stations (BS) at the top of the area.

Step 4: Assign id to each cluster. Each node and relay node will have their ids based on their cluster id.

Step 5: All the nodes, including relay nodes are given initial energies.

Step 6: Now join the nodes to their corresponding relay nodes which is located at the center of the cluster that node is in.
Step 7: Calculate the distance between the nodes and relay nodes, relay nodes and relay nodes.
Step 8: Now based on the distance and energies select the relay node with more energy and which is in the communication distance of node.
Step 9: Then join that node to that relay node. Like this groups are formed in every cluster.
Step 10: Now join the relay nodes with other relay nodes with more energy and which are in the communication range of that relay node.
Step 11: Then join the relay nodes. Now network is formed of relay nodes and normal nodes.
Step 12: Find the approximate Steiner points of the relay nodes using approximation algorithm.
Step 13: Now join the relay nodes to the corresponding Steiner points, new network is formed.
Step 14: Remove the direct connections between the relay nodes and put the relay- Steiner connections as it is.
Step 15: Find the minimum spanning tree of the network.

4.2   Second Scheme

First, the given plane is divided into many layers as per our requirements. Then, the communications are established. Within a cluster, a node can send the data to relay nodes which are in its communication range. Let the set of all relay nodes be $RN1$. Now, Algorithm 2 will join each node to its nearest relay node to form the groups. Now, these relay nodes within a single layer have to send the data to cluster heads. If the cluster head is closer they can directly communicate, otherwise a routing algorithm is used to connect the relay nodes and cluster head. Like this within each cluster, connections are formed. Then, the cluster heads will send the data to cluster coordinators in the above layer. Cluster coordinators will send the data to the cluster coordinators in the above layer and finally reach the base station.

Algorithm 2
Step 1: First deploy the nodes (N) and relay nodes (RN) in a fixed area.
Step 2: Divide the area into different or equal layers. Each layer is considered as a cluster.
Step 3: In each layer deploy the cluster heads (CH) one per layer, cluster coordinators (CCO) one per layer. And fix the base stations (BS) at the top of the area.

Step 4: Assign id to each cluster and initial energy to all the nodes and relay nodes.

Step 5: Now join the cluster heads to the cluster coordinators which are having id $Cid + 1$.

Step 6: Calculate the distance between the nodes and relay nodes, relay nodes and relay nodes, relay nodes and cluster heads.

Step 7: Now based on the distance and energies select the relay node with more energy and which is in the communication distance of node.

Step 8: Then join that node to that relay node. Like this groups are formed in every layer.

Step 9: Now within each layer join the relay nodes with other relay nodes with more energy and which are in the communication range of that relay node using a routing algorithm.

Step 10: Join the relay nodes which are closer to the cluster head directly, which forms the clusters in each layer.

Step 11: In each round algorithm will repeat from step 6 to 10.

4.3 Energy Expenditure Constraints

The various terms used with respect to the energy of each node is explained here. E_{Tx}, E_{Rx}: Energy consumption of a node for data transmission and receiving, respectively, E_{ele}: Energy consumption of radio electronics, ε_1, ε_2, ε_3, ε_4: Transmit amplifier of the normal node, relay node, cluster head, and CCO, $S_{(x,y)}$: Data rate for communication from node x to y, S_{max}: Maximum data rate, L: Data length of a packet. The following formulas are used for calculating the energy depleted while transmitting and receiving the data: $E_{Tx} = (E_{ele} + \varepsilon_1.d^2)L$, $E_{Rx} = L \times E_{ele}$.

Here, d is the distance between the node points. In each round of the Algorithm, the energy depleted will be deducted from the energies of nodes and relay nodes and iterated for various rounds. If the energy of node or relay node becomes null during the simulation, that node is omitted from selection. Relay nodes will have more initial energies than normal nodes, because they have a large workload.

5 Performance Evaluation and Results

5.1 Experimental Setup

The simulations are performed using MATLAB. The nodes are deployed in 100×100 m^2 area. The clusters are formed using **k means** clustering algorithm. Then, the relay nodes are deployed and the base stations are fixed at the top of the area. The ratio of relay nodes to the normal nodes may vary in each cluster. The energy of all the nodes is initially 0.5 J.

For the experiment, we used totally 1000 nodes, 10 relay nodes, 5 base stations. With these parameters and using the Algorithm 1, we formed a network of clusters. Simulation is run for 5000 rounds of data transmission and checked for different parameters each time by changing the number of nodes or relay nodes or number of clusters taken. The range of normal nodes and relay nodes r, R, respectively, is also varied. Depending on the parameters, network can be large or small. From this algorithm, we can easily scale the network into different layers even if the number of nodes and relay nodes is increased. The scheme used here shows that it is flexible with cluster topology and is energy-efficient.

5.2 Results

First scheme result: The parameters used in the experiments are as follows: Number of nodes = 1000, Number of relay nodes = 10, CH = Number of clusters = k = 10, number of base stations = 5, number of rounds = 5000. Figure 3 shows the number of dead nodes after several rounds for Algorithm 1.

Second Scheme Result

The following parameters are set up for the experiments. Number of nodes = 1000, number of relay nodes = 100, CH = number of clusters = k = 5, CCO = $k - 1$ = 4. Figure 4 shows the number of dead nodes after several rounds for Algorithm 2.

From Figs. 3, 4, and 5, we can see the number of alive nodes of LEACH and iLEACH protocols and our proposed algorithm. In Fig. 5, all nodes are dead in LEACH after 1000 rounds of data transmission and reception. In iLEACH after 1500 rounds, all the nodes are dead. But in the proposed algorithm, we can see that nodes are dead after 2250 rounds as depicted in Fig. 3. From Fig. 5, we can see that nodes are dead after 2500 rounds and still some nodes are alive after that.

5.3 Analysis of Results

From the above results, it is clear that the proposed algorithm performs better than LEACH and MODLEACH. Moreover in LEACH and MODLEACH for clustering, more time will be consumed in cluster formation. In the proposed algorithm, clusters are initially formed using *k means* clustering algorithm. In terms of energy, same number of nodes lasted for more rounds in the proposed algorithm which means energy is used very efficiently in forming clusters and data transmissions. In the proposed algorithm for forming the clusters, nodes will join the relay nodes having maximum energy, which results in slower energy depletion unlike other existing algorithms. The results indicate that the proposed algorithms have better performance, are more energy-efficient, and have more scalability for forming IoT.

Fig. 3 Plot of dead nodes versus number of rounds for Algorithm 1

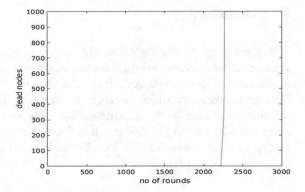

Fig. 4 Plot of dead nodes versus number of rounds for Algorithm 2

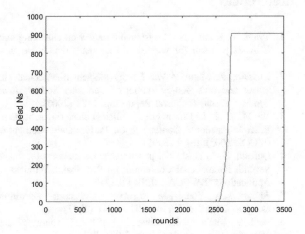

Fig. 5 Plot of dead nodes versus number of rounds for LEACH and iLEACH algorithms

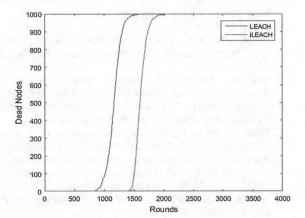

6   Conclusion

This paper presents two algorithms for achieving energy-efficient IoT. First, the structure of the network is defined for the deployment of IoT which has scalability features and makes it more extensible. After that, we developed algorithms, which form an IoT network which is energy-efficient and scalable. Efficient clustering techniques are used in the algorithms. Extensive simulations are performed and compared with existing algorithms. It is found that our proposed algorithms performed better than traditional WSN schemes in terms of network lifetime.

References

1. Tyagi, S., Kumar, N.: A systematic review on clustering and routing techniques based upon LEACH protocol for wireless sensor networks. J. Netw. Comput. Appl. **36**(2), 623–645 (2013)
2. Hussain, S., Matin, A.W.: Energy efficient hierarchical cluster-based routing for wireless sensor networks. Jodrey School of Computer Science Acadia University Wolfville, Nova Scotia, Canada, Technical Report, pp. 1–33 (2005)
3. Ye, M., et al.: EECS: an energy efficient clustering scheme in wireless sensor networks. 24th IEEE International Conference on Performance, Computing, and Communications, 2005 (IPCCC 2005). IEEE (2005)
4. Qureshi, T.N., et al.: On performance evaluation of variants of DEEC in WSNs. In: 2012 Seventh International Conference on Broadband, Wireless Computing, Communication and Applications (BWCCA). IEEE (2012)
5. Abbasi, A.A., Younis, M.: A survey on clustering algorithms for wireless sensor networks. Comput. Commun. **30**(14), 2826–2841 (2007)
6. Rani, S., Malhotra, J., Talwar, R.: EEICCP—energy efficient protocol for wireless sensor networks. Wirel. Sensor Netw. **5**(07), 127 (2013)
7. Norouzi, A., Babamir, F., Zaim, A.: A new clustering protocol for wireless sensor networks using genetic algorithm approach. Wirel. Sensor Netw. **3**(11), 362–370 (2011). https://doi.org/10.4236/wsn.2011.311042
8. Li, M.J., Ng, M.K., Cheung, Y.M., Huang, J.Z.: Agglomerative fuzzy k-means clustering algorithm with selection of number of clusters. IEEE Trans. Knowl. Data Eng. **20**(11), 1519–1534 (2008)
9. Rani, S., Ahmed, S.H.: Multi-hop routing in wireless sensor networks: an overview, taxonomy, and research challenges, p. 2015. Springer, Berlin, Germany (2015)
10. Rani, S., Malhotra, J., Talwar, R.: Energy efficient chain based cooperative routing protocol for WSN. Appl. Soft Comput. **35**, 386–397 (2015)
11. Huang, J., Meng, Y., Gong, X., Liu, Y., Duan, Q.: A novel deployment scheme for green internet of things. IEEE Internet Things J. **1**(2), 196–205 (2014)
12. Rani, S., Talwar, R., Malhotra, J., Ahmed, S.H., Sarkar, M., Song, H.: A novel scheme for an energy efficient internet of things based on wireless sensor networks. Sensors **15**(11), 28603–28626 (2015). http://www.mdpi.com/1424-8220/15/11/28603
13. Hwang, F.K., Richards, D.S., Winter, P.: The Steiner Tree Problem, vol. 53. Elsevier. eBook ISBN: 9780080867939

Minimizing the Total Range with Two Power Levels in Wireless Sensor Networks

Pushparaj Shetty D and M. Prasanna Lakshmi

Abstract Minimizing the total energy consumed by wireless sensor network (WSN) is a significant problem, since the sensor nodes are attached with a small battery of restricted capacity. In a WSN, any pair of sensor nodes must be able to communicate with each other in the network, so bidirectional connectivity of WSN is an important characteristic to be achieved. The range assignment problem in a WSN aims to assign transmission range to each sensor node of the network such that the specified connectivity constraints such as strong connectivity, k-connectivity are to be satisfied by the reduced network. Most sensors in recent days operate with discrete power levels. So, in this paper, we consider the range assignment problem with two power levels. Our aim is to assign each sensor node in the network with one of the available set of power levels such that the reduced topology is strongly connected and the total power consumption is minimized. The dual power assignment problem is well studied in the literature. We present an improved algorithm for dual power assignment problem in which the power levels are taken as input. Performance of the proposed algorithms is analyzed through extensive simulation. We establish the theoretical approximation ratio bound of the proposed algorithm for dual power assignment problem as 2. But, the simulation results indicate that the performance ratio is much less than 2.

Keywords Wireless sensor networks · Topology control problem
Power assignment · Graph algorithms · Approximation ratio · Minimum
spanning tree

P. Shetty D · M. P. Lakshmi (✉)
Department of Mathematical and Computational Sciences,
National Institute of Technology Karnataka, Surathkal 575025, India
e-mail: prasannasainitw@gmail.com

P. Shetty D
e-mail: prajshetty@nitk.edu.in

1   Introduction

A wireless sensor network (WSN) is a group of several sensor nodes, in which each node monitors and records several parameters such as temperature, humidity, pressure at various locations. There might be a discrete set of power levels available for assignment to each sensor node in a WSN. With this motivation of achieving bidirectional connectivity using discrete power levels, we study the range assignment problem with two power levels. Since the messages sent by a sensor node should be acknowledged by the receiver node, bidirectional links are given preference which also simplifies the routing protocols. Throughout the paper, the considered topology is bidirectional; therefore, we use undirected graph.

We model a WSN as an undirected graph, in which each vertex v represents a sensor node and the edge connecting any two nodes represents the communication link between that nodes. Based on the deployment, we also define a distance function, $d : V \times V \to R^+$ as $d(uv)$ = Euclidean distance between u and v. For a given set of vertices (or nodes) V, deployed in Euclidean plane with distance function $d : V \times V \to R^+$, **Range** of a vertex $v \in V$ is defined as:

$$R(V) = \max\{d(uv)|uv \in E\}. \tag{1}$$

2   Existing Results

Some WSNs assign same power level to every node and the details can be found in IEEE (1999) standard for Information technology, Wireless LAN medium access control (MAC) and physical layer (PHY) specifications. Lloyd et al. [8] have studied arbitrary power-level assignment for a WSN. In reality, discrete power levels are used, so k-power-level assignment problem was introduced which seeks a power assignment in which each node is assigned a power level from the k-power levels available for assignment satisfying certain connectivity constraints.

Rong et al. [10] have studied dual power management problem (DPMP) initially and proved its **NP**-completeness. In this paper, the authors proposed a 2-approximation algorithm, obtained lower and upper bounds for the optimal number of sensor nodes assigned with high power. The lower and upper bounds given by Rong et al. [10] are K and $2(K - 1)$ respectively, where K is the number of components the resultant network has after assigning low power to all the nodes. Carmi and Katz [3] proved that the dual power assignment problem is **NP**-hard using a different polynomial time reduction. Lam et al. [6] have studied dual power assignment problem with connectivity requirement as k-edge connectivity and proposed a 2-approximation algorithm [7].

Theorem 1 [10] *A feasible solution to the DPMP exists if and only if high power is at least maximum length of all the edges in Euclidean MST.*

Chen and Huang [4] have studied the problem of strong connectivity in multihop packet radio networks and proposed a 2-approximation algorithm which initially considers an undirected minimum spanning tree (MST), [5] and later, the bidirectional connectivity is established. Calinescu [2] proposed a greedy algorithm for dual power assignment problem which has the approximation ratio of 1.85. The problem of finding a 2-connected network that minimizes the total power consumption was studied by Panda and Shetty [9]. Sisodiya and Shetty [11] have studied the dual power assignment problem and proposed a heuristic named SCDPA of average approximation ratio 1.51 which is inferred from simulation results. Abu-Affash [1] studied dual power assignment problem using Hamiltonian cycle and proposed an algorithm of approximation ratio 1.571.

In this paper, we study dual power assignment problem under two cases: (i) the power levels are chosen from the given set and (ii) the power levels are taken as input. We also give the theoretical approximation ratio for this problem.

Rest of the paper is organized in the following way: In Sect. 3, we propose an algorithm for dual power assignment problem for which power levels are to be chosen from the given set. In Sect. 4, we present an algorithm for the same problem in which power levels are taken as input. Section 5 presents the results and discussion. Section 6 compares the proposed algorithm with existing (DPMP) [10]. Finally, Sect. 7 concludes the paper.

3 Dual Power Assignment Problem

The two power levels are denoted by H_p (high power) and L_p (low power) for the considered problem. The objective is to determine the two power levels and assign each node with any one of the two such that the induced subgraph is connected and the total power consumption of the network is minimized, and the problem is formulated as follows:

Problem: Dual power assignment—minimizing total power (DPA-MTP)
Instance: Set V of n nodes deployed in Euclidean plane and distance function $d : V \times V \to R^+$.
Objective: Determine the two power levels, H_p and L_p and assign each node with one of the two power levels such that the induced graph is connected and the total power is minimized.

3.1   Algorithm DPA-MTP

Let V be a set of nodes deployed in Euclidean plane with distance function d. Each edge joining two nodes u and v is associated with the Euclidean distance between u and v, i.e., $d(uv)$. Let T be the Euclidean MST of deployed nodes and each node is assigned a range as given in Eq. 1. Either low or high power is assigned to each node based on its corresponding range in spanning tree. To maintain the connectivity of graph, the minimum value of the high power H_p should be the at least length of an edge joining two farthest nodes in spanning tree (T). Therefore, we choose high power as follows:

$$H_p = \max\{R(v)|v \in V(T)\}. \tag{2}$$

Now, we determine power level L_p such that the induced graph is connected and the total power is minimized. Let the number of nodes assigned H_p and L_p be hno and lno, respectively. Then, we have $T_p = hno \cdot H_p + lno \cdot L_p$, here $hno + lno = n$, for n number of nodes. Next, we assign each node power level L_p if the range of that node is less than or equal to L_p in T and remaining all nodes are assigned H_p. Here we present pseudocode in Algorithm 1.

Algorithm 1: DPA-MTP
Input: Set of n nodes deployed and $d : V \times V \rightarrow R^+$.
Output: Dual power assignment minimizing the total power.
Step 1:   Find MST say T and assign range $\forall v \in V(T)$.
Step 2:   Let $H_p = \max\{R(v)|v \in V(T)\}$, $T_p = \infty$.
Step 3: $Lowp = \{d(uv)|u, v \in V, d(uv) < H_p\}$, $|Lowp| = count$.
Step 4:   For $i = 1$ to $count$ do
              $lno = hno = 0$, $L_p = Lowp(i)$
              For each $v \in V$ do
              If $(R(v) \leq L_p)$ $lno = lno+1$
              Else $hno = hno+1$
              End For
              If $(T_p > hno.H_p + lno.L_p)$

$$T_p = hno.H_p + lno.L_p, \; L_p = Lowp(i).$$

              End For
Step 5:   For each node $v \in V$
              If $(R(v) \leq L_p)$ $R(v) = L_p$
              Else $R(v) = L_p$.
              End For
Step 6:   Return H_p and L_p.

4 Variation of Dual Power Assignment Problem

In dual power assignment problem, the power levels are chosen, but in some cases, it might be required to assign power level from a given set. In that case, each node is assigned with one of the given two power levels such that connectivity is maintained and the total power is minimized. Minimizing the number of nodes assigned high power will minimize the total power in the WSN. Therefore, our objective is to minimize the number of high-power nodes and the problem is formulated as follows:

Problem: Variation of dual power assignment—minimizing the total power (VDPA-MTP).
Instance: A set V of n nodes deployed $d : V \times V \rightarrow R^+$, H_p and L_p.
Objective: Power assignment minimizing the number of high-power nodes such that the induced graph is connected.

4.1 Algorithm VDPA-MTP

The idea of the algorithm is as follows: First we assign low power to all the nodes that induces a spanning forest consisting of one or more components. Every node can communicate with all other nodes in its component. Now we find a vertex v which connects to maximum number of connected components, if it were assigned high power. Let that vertex be v from the component c. Let $v_1, v_2, \ldots, v_k$, be the vertices from the components $c_1, c_2, \ldots, c_k$, respectively, reaching the vertex v if they were assigned high power. Next, the vertices $v_1, v_2, \ldots, v_k$ and the vertex v are assigned high power. Now $c_1 \cup c_2 \cup \ldots \cup c_k$ will be the new component. We repeat this procedure until the resultant graph becomes single connected component. Based on this idea, we present the pseudocode for VDPA-MTP problem in Algorithm 2.

Algorithm 2: VDPA-MTP
Input: Set of n nodes deployed in Euclidean plane with the distance function $d : V \times V \rightarrow R^+$, H_p and L_p.
Output: Dual power assignment minimizing the total power.
Step 1: Find MST say T and $R(v) = L_p$, $\forall v \in V(T)$.
Step 2: Let C be the set of components at any instant.
Step 3: While $(|C| > 1)$ do
          Let u be the vertex reaching maximum number of components with high power and belongs to c.
          $R(u) = H_p$, $hno = hno + 1$.
          Let $c_1, c_2, \ldots, c_k$, be the components u reaches
          For $i = 1$ to k do

Find a vertex $w \in c_i$ that reaches vertex u with H_p
$R(w) = H_p$, $hno = hno + 1$, $T = T \cup \{uw\}$, $c = c \cup c_i$.
End For
End While
Step 4: Return T_p, hno.

Theorem 2 *Algorithm VDPA-MTP always results in a spanning tree of components.*

Proof Let T_c be the resultant graph on components obtained by VDPA-MTP algorithm, which is connected. Now to prove that T_c is a spanning tree, it is enough to prove that it is acyclic. If possible, let T_c be cyclic. When an edge is added to T_c, it is sure that its endpoints are on different components. So, deletion of any edge from T_c results into a disconnected graph. But, removal of an edge from a cycle does not disconnect the graph, which contradicts our assumption that T_c is cyclic.

Theorem 3 *VDPA-MTP is 2-approximation algorithm.*

Proof Let the VDPA-MTP algorithm results in a spanning tree of components $T_c = (V_c, E_c)$, each vertex $v_c \in V_c$ represents a component and each edge in E_c, connects two components. It is obvious that, to induce a connected graph, at least one vertex in each component should be assigned high power. If the degree of a vertex (component) $v_c \in V_c$ is one, only one vertex in that component is assigned high power. For any internal node $v_c \in V_c$ of degree m, the number of nodes assigned high power in that component is at least one and at most its degree, i.e., m. In worst case, for each vertex of T_c which is a component, the number of high-power nodes is its degree. We have $\sum {}^\circ(v_c) = 2|E_c| = 2(K - 1)$. Therefore, $K \leq hno \leq 2(K - 1)$, where K is the number of components obtained when each node is assigned low power initially.

5   Comparison with DPMP Algorithm

In this section, the proposed algorithm is compared with the existing, i.e., DPMP algorithm [10]. DPMP algorithm initially constructs a virtual graph G on components formed by assigning low power to all the nodes. Next, it randomly chooses a spanning tree T of G on components. In the spanning tree T, it chooses a leaf component and its adjacent component say C_u and C_v, respectively. If a pair of nodes i and j in C_u and C_v, respectively, can reach each other with high power, then i and j are assigned high power and C_u is deleted from T. This procedure is repeated until it remains with single connected component. Here the solution depends on the spanning tree it chooses. In the algorithm VDPA-MTP, we find a vertex which connects to maximum number of components with high power which results in more number of leaf components. In the leaf component, it is sufficient to assign high power to only one node. So, for all the leaf components, the number of connected components and the number of nodes assigned with high power will be the same.

6 Results and Analysis

We present the experimental results in this section, for which we randomly deployed set of nodes in Euclidean plane of 100×100. Every pair of nodes is assigned a positive real value, i.e., the Euclidean distance between the two nodes. We ran the algorithms DTP-MTP and VDPA-MTP for varying number of nodes.

Figure 1 shows low and high powers obtained by DPA-MTP algorithm for varying number of nodes. Without loss of generality, we considered integer values for power levels as input for implementation of VDPA-MTP algorithm. Figure 2 compares the number of components and the nodes with high power obtained in both the algorithms DPMP and VDPA-MTP for the same instance. It is observed that the number of nodes assigned with high power obtained by VDPA-MTP algorithms is less than or equal to that of DPMP algorithm [10]. The VDPA-MTP algorithm gives better results as compared to the established theoretical performance bound of 2.

We choose high power as given in Eq. 2 and for this value of high power; low power is varied from 5 to 80% of the high power. Figure 3 shows the result for $n = 100$, which depict that the ratio of low power to high power is inversely proportional to the number of components formed by assigning low power to all the nodes initially.

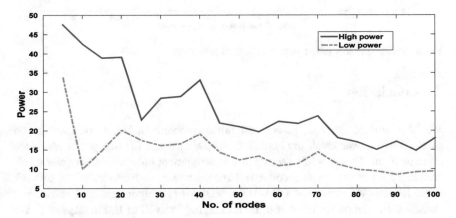

Fig. 1 High and low powers by DPA-MTP

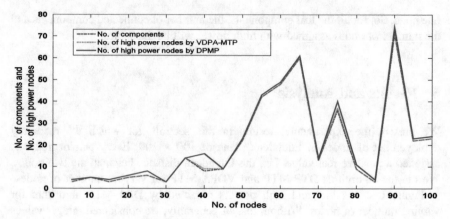

Fig. 2 Comparison between DPMP and VDPA-MTP

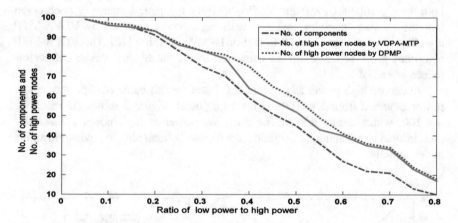

Fig. 3 Variation of low power from 5 to 80% of high power ($n = 100$)

7 Conclusion

We have studied the dual power assignment problem in WSN considering two cases (i) two power levels are chosen from given set and (ii) two power levels are taken as input. This problem seeks a power assignment only with two power levels such that the induced graph is connected and the power consumption of the network is minimized. We presented DPA-MTP, VDPA-MTP algorithms and carried out extensive simulation for number of nodes ranging from 10 to 100 in steps of 5. We established that the approximation ratio bound of the proposed algorithm VDPA-MTP is 2. But the simulation results indicate that the performance ratio is much less than 2. We believe that a careful theoretical analysis might result in a better approximation ratio.

References

1. Abu-Affash, A.K., Carmi, P., Tzur, A.P.: Dual power assignment via second hamiltonian cycle. J. Comput. Syst. Sci. **93**, 41–53 (2018)
2. Calinescu, G.: 1.85 approximation for min-power strong connectivity. CoRR abs/1205.3397 (2012)
3. Carmi, P., Katz, M.J.: Power assignment in radio networks with two power levels. Algorithmica **47**(2), 183–201 (2007)
4. Chen, W., Huang, N.: The strongly connecting problem on multihop packet radio networks. IEEE Trans. Commun. **37**(3), 293–295 (1989)
5. Cormen, T.H., Leiserson, C.E., Rivest, R.L., Stein, C.: Introduction to algorithms, 3 edn. MIT Press (2009)
6. Lam, N.X., Nguyen, T.N., An, M.K., Huynh, D.T.: Dual power assignment optimization for k-edge connectivity in WSNs. In: Proceedings of the 8th Annual IEEE Communications Society Conference on Sensor, Mesh and Ad Hoc Communications and Networks, SECON 2011, 27–30 June 2011, Salt Lake City, UT, USA, pp. 566–573 (2011)
7. Lam, N.X., Nguyen, T.N., An, M.K., Huynh, D.T.: Dual power assignment optimization and fault tolerance in wsns. J. Comb. Optim. **30**(1), 120–138 (2015)
8. Lloyd, E.L., Liu, R., Marathe, M.V., Ramanathan, R., Ravi, S.S.: Algorithmic aspects of topology control problems for ad hoc networks. MONET **10**(1–2), 19–34 (2005)
9. Panda, B.S., Shetty, D.P.: Minimum range assignment problem for two connectivity in wireless sensor networks. In: Distributed Computing and Internet Technology—10th International Conference, ICDCIT 2014, Bhubaneswar, India, 6–9 Feb 2014, pp. 122–133. Proceedings (2014)
10. Rong, Y., Choi, H., Choi, H.: Dual power management for network connectivity in wireless sensor networks. In: 18th International Parallel and Distributed Processing Symposium (IPDPS 2004), CD-ROM/Abstracts Proceedings, 26–30 Apr 2004, Santa Fe, New Mexico, USA (2004)
11. Sisodiya, N., Shetty, D.P.: Total power minimization using dual power assignment in wireless sensor networks. In: 2015 International Conference on Information Technology, ICIT 2015, Bhubaneswar, India, 21–23 Dec 2015, pp. 26–30 (2015). https://doi.org/10.1109/ICIT.2015.33

Part III
Electronics, Antenna Design

Comparative Study of Rectangular and E-Shaped Microstrip Patch Antenna Array for X-Band Applications

Alok Kumar Rastogi, Gazala Pravin and Shanu Sharma

Abstract This paper deals with the characterization of microstrip patch antenna array. These antennas are designed for X-band frequency applications. In the design process, SONNET software is utilized and antenna parameters such as current density, return loss, radiation pattern and voltage standing wave ratio (VSWR) are plotted. The rectangular and E-shaped patch antenna array is realized on 2×2 mm^2 Roger RT5880 sheet with thin layer of copper deposited over it, while the feed line is implemented using microstrip lines.

Keywords Microstrip patch antenna array · X-band frequency
Current density · Return loss · Voltage standing wave ratio

1 Introduction

In wireless communication technology, antenna is basically used to transmit and receive electromagnetic signals. By transmitting a signal into radio waves, the antenna transforms electric current into electromagnetic wave and vice versa by receiving [1].

Antenna plays an important role in today's telecommunication system. It is the heart of communication system as it is used to radiate as well as receive electromagnetic signals [2]. Today, main focus in communication is for circuits with reduced size, but improved performance is growing. So to reduce the size of circuit, antenna array is introduced and the basic and widely used patch antenna is microstrip. Microstrip patch antennas find its application in telemetry, radars, satellite, and GPS systems [3].

A. K. Rastogi (✉) · G. Pravin · S. Sharma
Department of Physics and Electronics, Institute for Excellence
in Higher Education, Bhopal 462016, India
e-mail: akrastogi_bpl@yahoo.com

© Springer Nature Singapore Pte Ltd. 2019
J. K. Mandal et al. (eds.), *Advanced Computing and Communication Technologies*,
Advances in Intelligent Systems and Computing 702,
https://doi.org/10.1007/978-981-13-0680-8_18

There are several feeding techniques such as coaxial probe, aperture coupling, proximity coupling, but here microstrip feeding is used. A sheet of dielectric substrate with very thin layer of copper layer deposited on both sides of the substrate is used for fabrication of microstrip patch antenna. One side acts as ground plane, whereas other metallization side can be of any shape such as rectangular, square, elliptical [4–6]. Demand for wideband, high-gain antenna is growing. So, single-element microstrip antennas are replaced by microstrip patch antenna arrays, which improve the performance over single-element microstrip antenna as well as E-shaped [7].

Rectangular as well as E-shaped microstrip patch antenna arrays were designed and simulated by using SONNET software. The proposed antenna arrays operate with center frequency approximately 10.9 GHz that supports at X-Band.

2 Geometry of Patch Antenna Array

The basic structure of rectangular patch antenna is shown in Fig. 1. In this method, RF power is fed directly to the radiating patch using microstrip line conducting element. This kind of feed arrangement has the advantage that the feed can be etched on the same substrate to provide a planar structure. 2D structure of rectangular and E-shaped patch antenna arrays is simulated on Roger RT5880 substrate with dielectric constant of 2.2 and loss tangent, tan δ = 0.0009, which are shown in Fig. 2a, b, respectively. The thickness of the substrate is 0.8923 mm. The size of patch length (L) is 8.15 mm, patch width (W) is 11.85 mm, and feed lines of rectangular and E-shaped patch antenna array are 22.1 and 19.25 mm, respectively, which is suitable for most satellite and radar application.

Fig. 1 Basic configuration of microstrip patch antenna

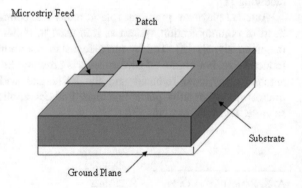

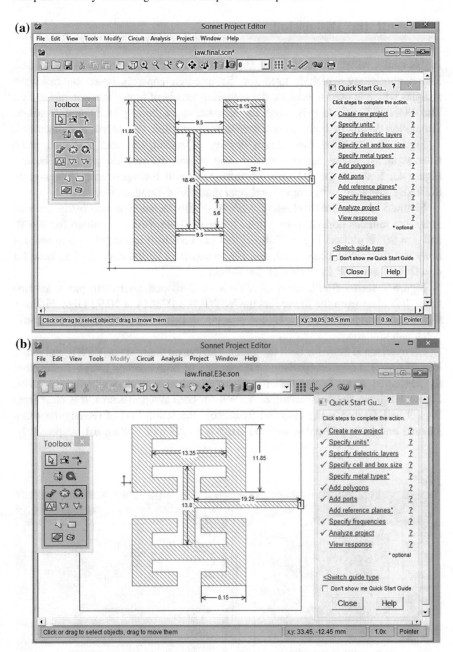

Fig. 2 2D Structure of **a** Rectangular patch antenna array, **b** E-shaped patch antenna array

3 Result and Discussion

In this study, wideband rectangular and E-shaped patch antenna arrays are studied. The methodology which is used for the research paper is based on full-wave analysis which is used to analyze the scattering matrix of both patch antenna arrays. The SONNET software is used for simulation of both patch antenna arrays. Simulation results of both antenna arrays are presented, compared, and discussed.

Figure 3a, b shows the current density of rectangular and E-shaped patch antenna array, where current density reflects how an antenna produces beams. Figure 4a, b shows the 3D view of rectangular and E-shaped array designed on SONNET software with single-band operation for X-Band applications.

Figure 5 shows Smith chart of VSWR of the rectangular microstrip antenna array that radiates normal to its patch surface, and the elevation pattern for $\varphi = 0°$ and $\varphi = 90°$ is plotted. Figure 5 shows that the VSWR of rectangular patch antenna array is 1.621608 at 10.86 GHz, and Fig. 6 shows radiation pattern of antenna for $\varphi = 0°$ and $\varphi = 90°$ in SONNET for E-field.

Figure 7 shows Smith chart of VSWR of E-shaped microstrip patch antenna array. It is clear from the figure that the VSWR is 1.023614 at 10.92 GHz. Figure 8 shows radiation pattern at $\varphi = 0°$ and $\varphi = 90°$ in E-field of the E-shaped microstrip patch antenna array.

Comparison is done for both the antenna arrays on the basis of VSWR and return loss. Figure 9 shows comparison of return loss of both antenna arrays. Return loss for E-shaped array is far better and improved than rectangular-shaped array. Figure 10 shows the comparison of VSWR for both antenna arrays. It is clear from Figs. 9 and 10 that at the resonant frequency, the return loss of rectangular patch antenna and E-shaped patch antenna array is −12.5 and −38.66 dB, respectively.

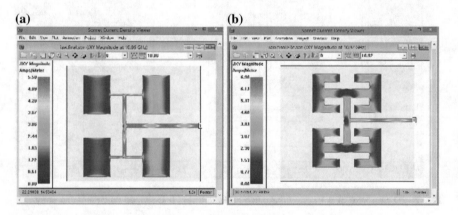

(a)   **(b)**

Fig. 3 Current density of **a** rectangular and **b** E-shaped patch antenna array

(a)   **(b)**

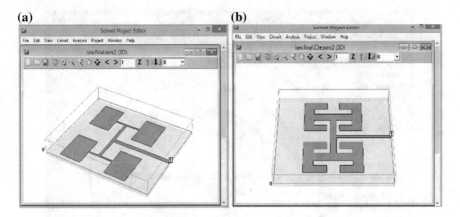

Fig. 4 3D view of **a** rectangular and **b** E-shaped patch antenna array

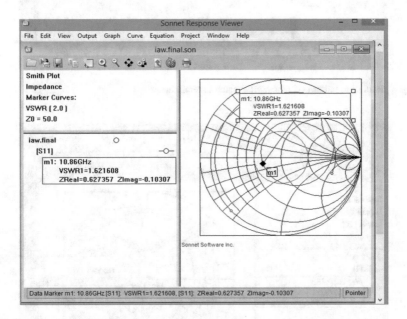

Fig. 5 Smith chart of VSWR of rectangular patch antenna array

Return loss of E-shaped antenna array is nearly 3 times that of rectangular patch antenna array; i.e., maximum power is reflected from E-shaped array. Also, VSWR is 1.66 and 1.024, respectively. VSWR is mismatch between antenna and feed line, which should be as small as possible. At resonant frequency, VSWR for E-shaped patch antenna array is closer to 1.

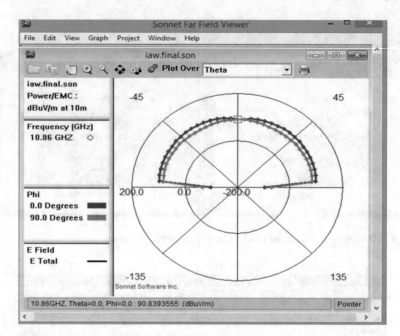

Fig. 6 Radiation pattern of rectangular patch antenna array

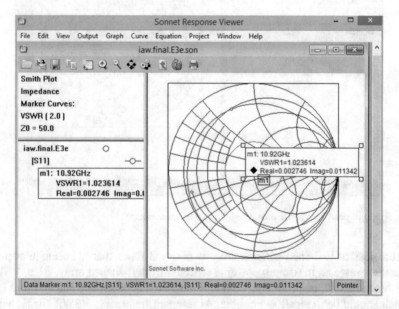

Fig. 7 VSWR of E-shaped patch antenna array

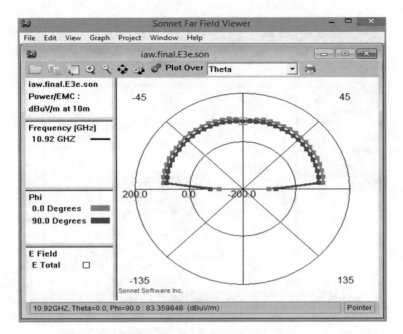

Fig. 8 Radiation patten of E-shaped patch antenna array

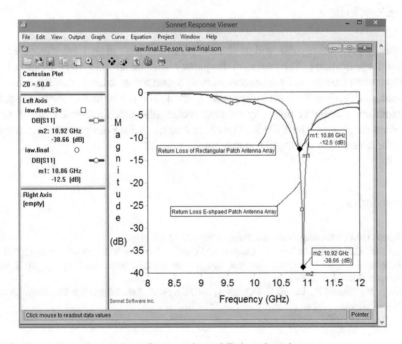

Fig. 9 Comparison of return loss of rectangular and E-shaped patch antenna array

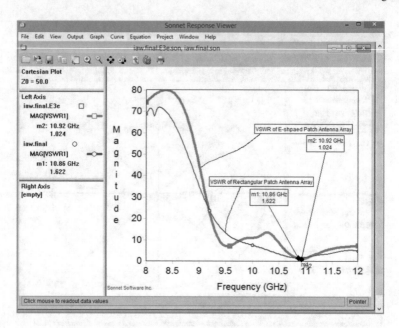

Fig. 10 Comparison of VSWR of rectangular and E-shaped patch antenna array

4 Conclusion

In this paper, rectangular and E-shaped patch antenna arrays are discussed for X-band applications. The resonant frequency was taken 10.9 GHz [approx.] for both antenna design, and parameters such as S-parameter, current density, radiation pattern, and VSWR for both arrays were plotted. The return loss of E-shaped patch antenna array is better than any existing rectangular patch antenna array at same resonant frequency. The VSWR at resonant frequency for E-shaped antenna array is also far better than rectangular patch antenna array.

References

1. Pozar, D.M.: Microstrip antenna. Proc. IEEE **80**, 77–81 (1992)
2. Balanis, C.A.: Antenna Theory Analysis and Design, 3rd edn. Wiley, New Jersey (2005)
3. Beenamole, K.S.: Microstrip Antenna Designs for Radar Applications, pp. 84–86. DRDO Science Spectrum (2009)
4. Garg, R., Bhatia, P., Bahal, I., Ittipiboon, A.: Microstrip Antenna Design Handbook. Artech House (2001)

5. Rastogi, A.K., Pravin, G.: Design and simulation of H and E-shaped microstrip patch antenna for S band communication. Int. J. Eng. Res. Technol. (IJERT) **5**, 349–353 (2016)
6. Carver, K.R., Mink, J.W.: Microstrip antenna technology. IEEE Trans. Antennas Propag. **29**, 2–24 (1981)
7. Sharma, N., Jain, B., Singla, P., Prasad, R.R.: Rectangular patch micro strip antenna: a survey. Int. Adv. Res. J. Sci. Eng. Technol. **1**, 144–147 (2014)

An Archimedean Spiral-Shaped Frequency-Selective Defected Structure for Narrowband High Q Applications

Somdottta Roy Choudhury, Aditi Sengupta, Susanta Kumar Parui and Santanu Das

Abstract An Archimedean spiral-shaped frequency-selective defected structure is proposed in this paper. This type of structure provides high Q band reject performance with a very small dimension of $0.02\lambda_0 \times 0.02\lambda_0$. The structure is simulated and mathematically analyzed, and equivalent circuit representation is also made. It is observed that the investigated DGS unit provides a stopband centered at 5.1 GHz. The sharpness factor is very high almost greater than 697 and 398 dB/GHz for the lower and upper side of the stopband. The 20 dB rejection bandwidth is 0.27%. The loaded Q factor for the above dimensions is 46.5. The proposed structure is measured and compared with the simulated data.

Keywords High Q factor · Band rejection · Archimedean spiral shape

1 Introduction

Frequency-selective structure is 1D or 2D periodic structures which are widely investigated for decades [1, 2]. Having the frequency-selective property, these structures are applied in communication systems. Theoretically, these are planar and infinite structures; however, for the practical applications, they are usually placed in limited space with limited number of elements [1, 2]. Defected frequency-selective structure at the ground plane can produce band rejection in certain frequency bands [3–6]. These structures usually add an extra lumped inductance and capacitance to the microstrip line, which are connected as a parallel resonant circuit in series with transmission lines at both of its ends [7]. Here in this

S. R. Choudhury (✉)
Ramrao Adik Institute of Technology, Nerul, Navi Mumbai 400706, India
e-mail: somdottaroychoudhury@gmail.com

A. Sengupta
Gurunanak Institute of Technology, Panihati, Sodepur 700114, India

S. K. Parui · S. Das
Indian Institute of Enginering Science and Technology, Howrah 711103, India

© Springer Nature Singapore Pte Ltd. 2019
J. K. Mandal et al. (eds.), *Advanced Computing and Communication Technologies*,
Advances in Intelligent Systems and Computing 702,
https://doi.org/10.1007/978-981-13-0680-8_19

paper, an Archimedean spiral-shaped defected frequency-selective structure is placed at the ground plane. The unit cell of the structure is simulated, mathematically analyzed, and measured by vector network analyzer. This structure yields a very narrowband high Q bandstop response which is designed for the WLAN (5 GHz) application. The structure is practically realized, and measured result is in good agreement with simulated result.

2 Design

Figure 1 shows that the schematic diagram of the frequency-selective structure consists of Archimedean spiral-shaped slot etched off under the microstrip line. The substrate with a dielectric constant of 3.2 and thickness of 0.79 mm is considered for the microstrip line. The width (W) of the microstrip line is obtained as 1.92 mm corresponding to 50 Ω characteristic impedance. In the same figure, the 'o' is the origin of the x–x'- and y–y'-axes. The different dimensions of Archimedean spiral-shaped DGS have been taken as $a = 0.3$ mm, $s = 0.1$ mm, $d = 0.1$ mm. The DGS has overall cut out size of the proposed structure is 1.9×2 mm^2.

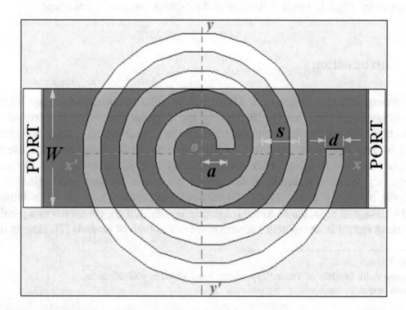

Fig. 1 Layout of the Archimedean spiral-shaped frequency-selective structure

2.1 Simulated Result

From the MoM-based IE3D-simulated S-parameters as shown in Fig. 2, it is observed that the investigated DGS unit provides a stopband with 3 dB cutoff frequency at 5.06 and 5.17 GHz with the pole frequency is at 5.1 GHz. The sharpness factor is very high almost greater than 697 and 398 dB/GHz for the lower and upper side of the stopband. The 20 dB rejection bandwidth is 0.27%. The loaded Q factor for the above dimensions is 46.5.

2.2 Mathematical Analysis

Analyzing the proposed structure, it clearly observed that the length of the arm of the Archimedean spiral has the effect on the resonant frequency (f_r). This length can be calculated as by the given equation (Eq. 1) below.

Displayed equations are centered and set on a separate line.

$$L = \int_{m}^{n} \sqrt{(r^2 + b^2)} d\theta \tag{1}$$

where

$$r = a + b\theta \tag{2}$$

In the above equation, m and n are the upper and lower limit of the integration where n = total number of turns × 2. In Eq. (2), a = distance of the start point of the spiral from the origin 'o', and b = (distance between each arm × total no. of

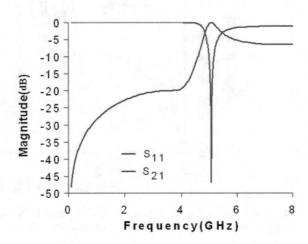

Fig. 2 Simulated S-parameter response of the proposed structure

Table 1 Comparison of the calculated and simulated resonant frequencies 'f_r' for different values of 'L'

Length of arm 'L' (mm)	Calculated 'f_r' (GHz)	Simulated 'f_r' (GHz)
18.7	3.75	3.79
23.5	3.01	3.20
25.2	2.78	2.75

turn)/2. Since, the length (L) is related to the guided wavelength as $L = 3\lambda_g/8$ (inductive load) and the corresponding resonant frequency 'f_r' is obtained by Eq. (3) below. Table 1 gives a list of calculated and simulated resonant frequencies for different 'L' values, and it is found that there are great resemblances between the calculated and simulated results (Fig. 3).

$$f_r = \frac{c}{2.667L\sqrt{\varepsilon_{\text{eff}}}} \tag{3}$$

The change of the length 'L' of the spiral-shaped DGS causes change in cutoff and transmission zero frequency which is depicted in Fig. 4b. The slot width 'd' and strip width 's' have an effect on the loaded Q factor. If 'd' and 's' decreased simultaneously, then the self and mutual inductions are increased, and as a result, the Q factor is also increased. Table 2 shows the variation of Q factor with the simultaneous change in 'd' and 's' for the same resonant frequency of 5.1 GHz. Therefore, the proposed DGS structure can be applied to very narrowband highly selective filter application.

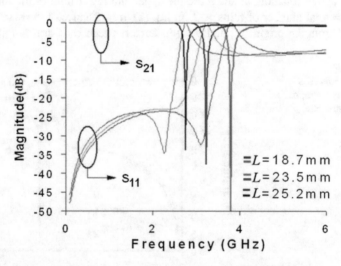

Fig. 3 EM-simulated S-parameter response of the spiral DGS filter showing that variation of the length 'L' causes change in upper cutoff and upper transmission zero

Fig. 4 Variation of the length 'L' with the change in upper cutoff and upper transmission zero

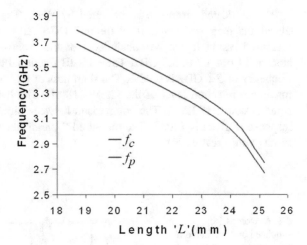

Table 2 Variation of loaded Q factor with slot width 'd'

Resonant frequency 'f_r' (GHz)	Slot width 'g' (mm)	Strip width 's' (mm)	Loaded Q factor
5.1	0.3	0.3	12.75
5.1	0.2	0.2	25.5
5.1	0.1	0.1	46.5

3 Measured Result

The proposed DGS prototype is fabricated on a PTFE substrate with dielectric constant of 3.2 and loss tangent of 0.002. Due to the limitation of the process used for fabricating proposed prototype, the dimension of less than 0.3 mm is not possible to realize in the laboratory. Therefore, another structure with dimensions: $a = 0.3$ mm, $s = 0.1$ mm, $d = 0.1$ mm is fabricated. The photographic view of the structure is given in Fig. 5.

Fig. 5 Fabricated prototype of the proposed structure

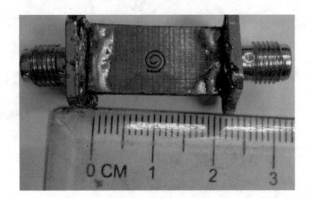

The fabricated prototype is measured by VNA. The comparison between simulated and measured response of the unit DGS cell is shown in Fig. 6. From the measured result, it can be concluded that the proposed filter exhibits a narrow stopband from 4.9 to 5.3 GHz. The −15 dB fractional bandwidth (FBW) at center frequency of 5.1 GHz is 0.98%. The sharpness of the stopband is 75 dB/GHz. The maximum rejection level of the filter is 18 dB. The loaded Q factor for the measured structure is 12.75. The comparison between simulated and measured phase response is given in Fig. 7. The measured S-parameter plots comply the simulated results with great extent.

Fig. 6 Comparison of simulated and measured S-parameter response of the proposed structure

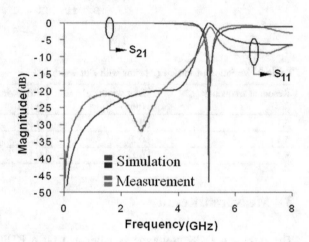

Fig. 7 Plot of measured phase response of the proposed structure

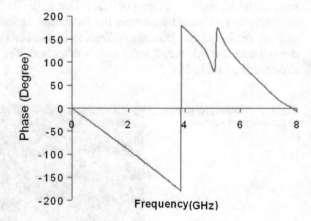

4   Conclusion

Here in this paper, an Archimedean spiral type of defected frequency-selective structure and its characteristics are studied. This defected structure provides a narrow stopband performance and provides highly sharp roll of having a sharpness factor of 697 dB/GHz and a very narrowband bandstop response. This kind of filter is very compact having cross-sectional area of $0.02\lambda_0 \times 0.02\lambda_0$ and can be used as notch band filter for eliminating unlicensed bands. The loaded Q factor is quite high and is given by 46.5. However, change in coupling gap required for feasible fabrication process causes a great variation of the loaded Q factor also.

References

1. Rashid, A.K., Shen, Z.: A novel band-reject frequency selective surface with pseudo-elliptic response. IEEE Trans. IEEE Antennas Propag. **58**(4), 1220–1226 (2010)
2. Rashid, A.K., Li, B., Shen, Z.: An overview of three-dimensional frequency-selective structures. IEEE Trans. IEEE Antennas Propag. Mag. **56**(3), 43–67 (2014)
3. Kim, C.S., Park, J.S., Ahn, D., Lim, J.B.: A novel 1-D periodic defected ground structure for planar circuits. IEEE Microw. Guid. Wave Lett. **10**(4), 131–133 (2000)
4. Ahn, D., Park, J.S., Kim, C.S., Kim, J., Qian, Y., Itoh, T.: A design of the low-pass filter using the novel microstrip defected ground structure. IEEE Trans. Microw. Theory Tech. **49**(1), 86–93 (2001)
5. Park, J.-S., Yun, J.-S., Ahn, D.: A design of the novel coupled-line bandpass filter using defected ground structure with wide stopband performance. IEEE Trans. Microw. Theory Tech. **50**(9), 2037–2043 (2002)
6. Park, J.-S., Kim, J.-H., Lee, J.-H., et al.: A novel equivalent circuit and modeling method for defected ground structure and its application to optimization of a DGS lowpass filter. IEEE MTT-S Int. Microw. Symp. Dig. **1**(4), 417–420 (2002)
7. Huang, S.Y. Lee, Y.-H.: A compact e-shaped patterned ground structure and its applications to tunable bandstop resonator. IEEE Trans. Microw. Theory Tech. **57**(3), 657–666 (2009)

Challenge-Response Pair (CRP) Generator Using Schmitt Trigger Physical Unclonable Function

Abhishek Kumar and Ravi Shankar Mishra

Abstract Physically unclonable function (PUF), a hardware security module, generates a unique response for each challenge by embodying a hardware feature into the computation. In the proposed Schmitt trigger(ST)-based PUF, the response is a function of circuit-based two different feature; triggering nature of ST generates a defined saturation level for changing input, and delay rises through cascading structure of STs. Multi-feature-based PUF sustains better unclonability and retains average reliability 99.725% with temperature variation, 99.86 with supply variation, and average uniqueness 47%. Another part of work is to enhance the response rate; rising and falling input sample individually generates five responses for rising and five responses for falling input. The proposed ST-PUF can generate ten numbers of 32-bit responses in one input period, and it enhances response generation rate by 10 times.

Keywords PUF · Schmitt trigger · CRP · Hardware security

1 Introduction

While communicating/transacting with the real world, we put a lot of trust in other person or computer system. We ensure that our actions and data must safe, are they really secure? There are different algorithm and protocol like AES, DES, SHA, RSA to answer this question. But software-based security mechanism is obsolete nowadays; security must arise from hardware root of trust. In digital world, security arises with 'password'; the system must be authenticated with the actual one. Each character of the password are encrypted with a secret key; derived from a random

A. Kumar (✉)
School of Electronics Engineering, Lovely Professional University,
Phagwara 144411, India
e-mail: abkvjti@gmail.com

R. S. Mishra
Symbiosis University of Applied Sciences, Indore 453112, India

© Springer Nature Singapore Pte Ltd. 2019
J. K. Mandal et al. (eds.), *Advanced Computing and Communication Technologies*,
Advances in Intelligent Systems and Computing 702,
https://doi.org/10.1007/978-981-13-0680-8_20

Fig. 1 PUF as
challenge-response pair
(CRP) generator

number and stored in memory usually EEPROM. The requirement of memory complicates the system and comes with cost; a memory constantly leaks the side channel information like leakage power, heat, radiation, timing etc. An adversary can utilize leakage current from memory to predict the stored data. A cryptosystem requires a random secret key originated from hardware mechanism. Security system requires a function that can generate a random and unique response every time for each challenge. Physical unclonable function (PUF) is a hardware security module preferred for secured authentication and cryptographic secret key generation [1–3]. PUF eliminates the memory; can generate the response (secret) runtime, it simplifies the complexity level and reduces cost. A PUF accepts input challenge that includes the intrinsic feature of system or circuits to generate the response as shown in Fig. 1. Silicon IC based PUF are popular; due to its manufacturing variation leads to high randomness in responses. The expectation of PUFs are (a) response must different for same challenge applied to multiple PUF, (b) unique response to each challenge, (c) similar response under different operating environment. All electronics properties variable in nature and settle with a stable state can use to turn into PUF.

2 Schmitt Trigger Architectures

2.1 Schmitt Trigger

Figure 2 presents a Schmitt trigger circuit that contains double transistor inverter by NM1, PM3 and NM0, PM4. The threshold voltage of NM1, PM3 is higher than NM0, PM4 provide inverting output swing after a finite delay, feedback transistor PM2 and NM2 provides hysteresis. For input voltage (Vin < Vth), NM0, NM1 in OFF state and PM3, PM4 in ON state keep output at a high level NM2 will be ON. When Vin approaches Vth of NM0; it will be ON and NM1 remain OFF since the output is high at this time NM2 will be ON. Output node pulls down by NM0 and pulls up by NM2, and the output stays at high output level. When Vin approaches Vth of NM1, it turn on pulls down out level to low state yield lower switching point for rising input. Similarly for high input NM0, NM1 in ON state and PM3, PM4 in OFF state provide low output keep PM2 ON. When Vin falls behind threshold of PM4, PM4 turns ON and PM3 remains OFF, PM4 tries to pull up, and PM2 pulls down the output node maintains at a low level. When Vin fall to the threshold voltage of PM3; it turn on pull up the output level to high yield high switching point for falling input. The difference between two switching point is known as hysteresis.

Fig. 2 Schmitt trigger [14]

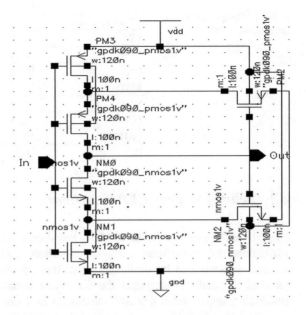

In this work, eight different type of Schmitt trigger circuit has been used to generate different delay and hysteresis discussed in [4–7]. Since series topology of MOS increase delay and parallel topology reduces delay, by adding PMOS in series (NOR_ST) maximizes the delay. SAH_ST, a self-adjusting Schmitt trigger, is provided a stack or MOS in series of feedback transistor; it adjusts the amount of feedback responsive to a supply voltage level so as to maintain or increase the hysteresis. STD_ST is a compensated Schmitt trigger circuit for providing a monotonic hysteresis response, and the feedback contains a plurality of transistors for providing a hysteresis response at higher supply voltage followed by an inverter. In a compensated ST, switching points of this circuit are more difficult to predict. SOI_ST have controllable hysteresis feature, contain a stack of PMOS and NMOS in feedback path. 2 pair of feedback transistor provides output with successive adjustment of upper and lower trigger edge. HEX_ST contains ST, latch, and inverter in cascade; latch enlarges the hysteresis region and delay. DT_ST is abbreviation for dynamic threshold ST, it is the cascaded structure of two inverting stage, the output of first and second stage is coupled and acts as third stage input, and third stage adjusts to provide a feedback voltage to second stage. Here, third stage dynamically adjusts the threshold voltage or trigger point. LADDER_ST is an extended form of SOI_ST, and it contains four NMOS and PMOS in series and provides three junction points to feedback transistor. Figure 4 presents 4 PUF (each PUF contains 2 parallel row of schmitt trigger controlled by multiplexer) acting on a traingular input, each ST sample the input at different trigger voltage yield different rising and falling edge. Delay and hysteresis value of each ST have been listed in Table 1.

Table 1 Comparison of ST architecture

ST architecture	Hysteresis	Delay (ps)
ST	0.22	205
NOR	0.04	110
HEX	0.355	488
STD	0.245	337
SAH	0.06	130
SOI	0.55	302
LADDER	0.25	392
DT	0.37	531

2.2 Schmitt Trigger CRP Generator

ST_PUF architecture is similar to MUX-based delay PUF discussed in [1, 3, 8, 9], where delay element is Schmitt trigger and switching action of multiplexer between the path is controlled by challenge input. Input challenge applied to a selection input of the multiplexer, select the ST which sample rising-falling input wave situated on upper and lower row mentioned in Table 2. Each row of ST_ PUF contains two delay paths which sample the variable input at unique trigger voltage, at the end input to arbiter is the cumulative effect of delay and trigger. Row0 of ST_PUF triggers the input signal NOR, SOI, STD, and LADDER-based ST architecture for the challenge '1010'. The arbiter at the end decides which signal is faster as shown in Fig. 3. Arbiter response is high if the upper path is faster than lower path else zero. Each row of ST_PUF can generate a 1-bit response, in order to have multi-bit response parallel n-number of ST-PUF populate in parallel.

Table 2 Combination of Schmitt trigger in PUF row

PUF0	PUF1	PUF2	PUF3	Output bit
STD	SOI	ST	LADDER	Bit0
NOR	SAH	STD	SAH	
LADDER	STD	NOR	SAH	Bit1
NOR	NOR	LADDER	STD	
STD	STD	LADDER	LADDER	Bit2
HEX	HEX	STD	NOR	
NOR	HEX	STD	NOR	Bit3
LADDER	LADDER	SAH	ST	
STD	STD	NOR	ST	Bit4
LADDER	SAH	STD	STD	
SAH	STD	HEX	STD	Bit5
STD	ST	SOI	HEX	
HEX	HEX	SOI	LADDER	Bit6
LADDER	SOI	ST	LADDER	
ST	SAH	LADDER	LADDER	Bit7
SOI	ST	SAH	SAH	

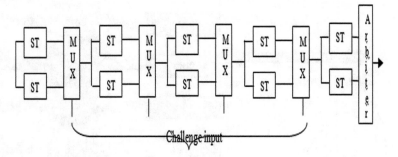

Fig. 3 Schmitt trigger PUF

$$\text{Response} = f(\text{challenge, delay \& trigger voltage}) \qquad (1)$$

The proposed work is implemented with cadence virtuoso CMOS 90 nm technology, and simulation result is verified with cadence spectre simulator. Architectural variation into PUF is introduced by different ST for the top and bottom path of row. Proposed PUF consists of eight Schmitt trigger circuits acting on a common signal passing through eight parallel paths possessing different delay and different triggering interval (hysteresis). Arrival time to the arbiter of each path is different; arbiter generates an 8-bit parallel response. Each PUF produces 8-bit parallel output PUF0 (D0-D7), PUF1 (D7-D15), PUF2 (D16-D23), and PUF3 (D24-D31). Table 2 represents the combination of Schmitt triggers into each row. First- and second-row outputs fed into arbiter produce bit0 of each PUF0, and similarly, third and fourth row produce bit1 and so on. For the simulation study 32-bit response for all 16 challenges is collected. The same design can extend to 8, 16, or higher number of stages to generate more response in parallel.

3 Schmitt Trigger PUF

3.1 Challenge-Response Generation Scheme

Figure 4 presents the proposed circuit of challenge-response generation; 4-bit challenge is mapped onto 32-bit responses. Triangular input is applied to 4 PUF acting concurrently. PUF output response followed by 8-DFF working on PIPO mode operates at the clock input. Triangular input period 2000 ps is sampled 10 times by clock generator time period 200 ps, clock select a small width of slope from the input. A small width of a variable input is sufficient to trigger and generate a saturation level since threshold voltages of STs are very small. Each PUF samples the input 10 times with clock frequency of 200 ps. Each ST triggers during this small change and generates different arrival time to arbiter; DFF enables for this interval and stores the response. Figure 5 shows the transient simulation response

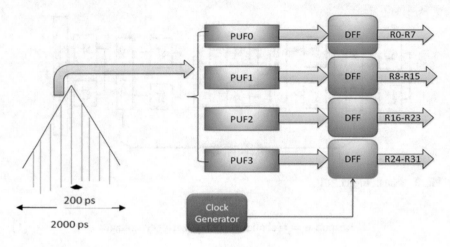

Fig. 4 CRP generation scheme

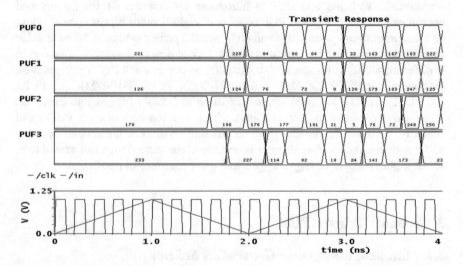

Fig. 5 Transient response of ST_PUF

during first clock input–output of PUF0, PUF1, PUF2, PUF3 is 220,124, 186, 227, respectively, and so on. Rising and falling input samples five times. Ten responses can be obtained from one input complete system throughput enhanced by 10 times.

3.2 Result and Analysis

Apart from the area, power, and delay, every PUF designs metrics are evaluated with inter- and intra-chip variation [10, 11]. PUF outputs are unique (for security), reproducible (for reliability), and uniformity distributed (for the unbiased result). Hamming distance (HD) model is most preferred one for PUF evaluation. HD among two digital responses refers to the minimum number of bits, which has to be changed to make the digital bits identical.

3.2.1 Inter-PUF Variation

Inter-PUF variation is a measure of a number of bit differences between responses of two PUF for the same challenge. On applying the same challenge to two different PUF designs, ideally, the Hamming distance between their responses should differ 50% of total responses bits [8, 9].

$$\text{Inter HD} = \frac{2}{K(K-1)} \sum_{i=1}^{K-1} \sum_{K=i+1}^{K} \frac{\text{HD}(R_i, R_j)}{n} \times 100\% \tag{2}$$

3.2.2 Intra-PUF Variation

Intra PUF variation is a measure of separation of response bit generated from a single PUF for 1-bit change in the input. On flipping a bit in the challenge to a PUF, ideally, the hamming distance between the responses should differ 50% of total responses bits [9, 12].

$$\text{Intra HD} = \frac{1}{K} \sum_{i=1}^{K} \frac{\text{HD}(R_{i1}, R_{i2})}{n} \times 100\% \tag{3}$$

3.2.3 Uniformity

Uniformity defines how uniform the proportion of '1's and '0's in the response bits of a PUF. Ideally, this value should be 50%.

$$\text{Uniformity} = \frac{1}{n} \sum_{j=1}^{n} R_{i,j} \times 100\% \tag{4}$$

Figure 6 presents the uniformity of Schmitt trigger PUF at sampling interval rate of 200 ps; uniformity falls with starting time and rises for falling slope of input. For

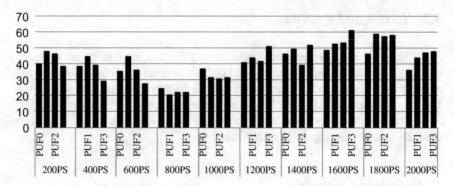

Fig. 6 Uniformity of ST-PUF

seventh sampling, clock uniformity is close to the ideal value. Average value of uniformity for all PUS ranges between 20 and 60.93%.

3.2.4 Bit Aliasing

Bit aliasing is a measure of biasness of the bit in the responses, few bits of response are stuck to '0' if biased to zero or stuck to '1' is biased to one. PUF provides a similar response for different PUF; it gives an exclusive hint to the attacker to guess the response. If bit aliasing happens, then different chips will produce similar responses. Consequently, an attacker can easily guess the response. Bit aliasing for the ith bit of a PUF across K different chips for a challenge C is estimated as

$$\text{Bit Alising} = \frac{1}{k} \sum_{j=1}^{K} R_{i,j} \times 100\%. \tag{5}$$

Figure 7 presents the bit aliasing metrics of proposed ST-PUF, its ideal value is 50%, and it can observe that bit1 of PUS is more biased and bit4 is less biased. Rising input is more biased towards '0' that produces stuck-a-0 fault, while falling input is biased towards '1' that produces a stuck-a-1 fault.

3.2.5 Uniqueness

Uniqueness is a measure of uncorrelated response for the similar challenge overall PUF instances [8, 13]. The response of a PUF should unique to each challenge, for same challenge-response generated by different PUF must be different. Its ideal value should be 50% of response bit; uniqueness is measured as the average value

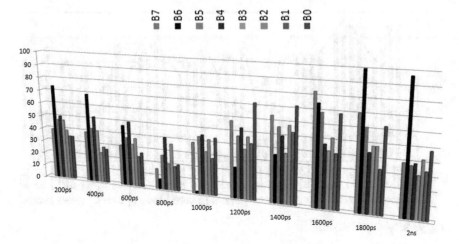

Fig. 7 Bit aliasing of ST-PUF

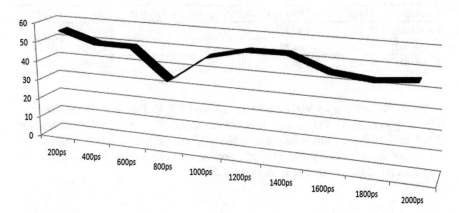

Fig. 8 Uniqueness of ST-PUF

of inter-PUF variation. Figure 8 shows that uniqueness falls to 34% for fourth clock interval; it touches to an ideal value falling input and remains close to 50%. Uniqueness ranges from 34 to 55.5%.

3.2.6 Reliability

The reliability of a PUF circuit refers to its ability to produce the same response for a given challenge over varying environmental conditions like temperature. It is observed that reliability falls at high-temperature value and higher supply due to process variation. PUF response should be stable with temperature and supply variation. In this work, four PUF response is analysed under temperature variation

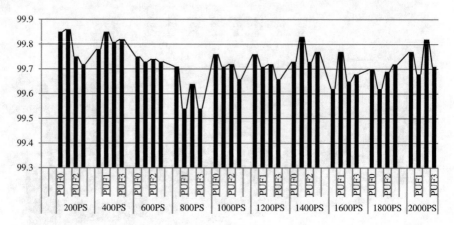

Fig. 9 Reliability of ST-PUF

Table 3 Comparison of ST_PUF

Parameter	Ideal (%)	Proposed ST_PUF (average value) (%)	[10]	[9] (%)
Uniformity	50	40.5	–	30.26
Uniqueness	50	44.75	42.8%	67.44
Reliability	100	99.7	80.6%	98.01

of 20–80 °C and supply variation of 1, 1.2, and 1.5 V. Figure 9 shows the reliability of proposed ST-PUF over a temperature range of 20–80 °C; the reliability of PUF1 and PUF3 attains minimum value 99.54% at fourth clock interval and maximum value by PUF0 and PUF1 99.86% at first clock interval. Comparison of proposed ST_PUF parameter is listed in Table 3.

4 Conclusion

The proposed Schmitt trigger PUF presents a novel scheme to generate challenge-response pair based on delay and trigger voltage. This is first time response which is computed by embodying multi-electronics property which significantly improves average reliability to 99.86% and average uniqueness 47%. Triangular input sample n-time by adjusting the period of the clock pulse can generate n-response for a challenge input; enhance the response rate n-times. The proposed PUF stands a competitive candidate for a large number of challenge-response pair generators.

References

1. Suh, G.E., Devadas, S.: Physically unclonable functions for device authentication and secret key generation. In: Proceedings of the ACM/IEEE Series Automation Conference, pp. 9–14 (2007)
2. Kumar, S., Guajardo, J., Maes, R., Schrijen, G.J., Tuyls, P.: The butterfly PUF: protecting IP on Every FPGA. In: Proceedings of the IEEE International Workshop on Hardware-Oriented Security and Trust, pp. 67–70 (2008)
3. Devadas, S., et al.: Design and implementation of PUF-based unclonable RFID IC for anti-counterfeiting and security applications. In: Proceedings of the IEEE International Conference RFID, pp. 58–64 (2008)
4. Chuang, et al.: SOI CMOS Schmitt trigger circuits with controllable hysteresis. US Patents 6,441,663-B1 (2002)
5. Chauhan, et al.: Compensated Schmitt trigger circuit for providing monolithic hysteresis response. US Patents 0017482-A (2006)
6. Barlow: Self adjusting Schmitt trigger. US Patents 7,167,032-B1 (2007)
7. Arith, F., et al.: Low voltage CMOS Schmitt trigger in 0.18 μm technology. IOSR J. Eng. **3** (3), 08–15 (2013)
8. Rostami, M., et al.: A primer on hardware security: models, methods, and metrics. Proc. IEEE **102**(8), 1283–1295 (2014)
9. Pegu, R., Audio, R.: Design and analysis of MUX based physical unclonable function. IEEE Trans. Comput.-Aided Design Integr. Circ. Syst. **33**(5), 649–662 (2015)
10. Maita, A., et al.: Physical unclonable function and true number generator: a compact and scalable implementation. In: Proceedings of the 19th ACM Great Lakes Symposium on VLSI, pp. 425–428 (2009)
11. Herder, C., et al.: Physically unclonable function and application: a tutorial. Proc. IEEE **102** (8), 1126–1141 (2014)
12. Cherkaoui, A., et al.: Evaluation and optimization of physical unclonable functions based on transient effect ring oscillators. IEEE Trans. Inf. Forensic Secur. 1291–1305 (2016)
13. Ozturk, E., Hammouri, G., Sunar, B.: Physically unclonable function with tristate buffers. In: Proceedings of the IEEE International Symposium on Circuits and Systems, pp. 3194–3197 (2008)
14. Filanocsky, I.M., Bakes, H.: CMOS Schmitt trigger design. IEEE Trans. Circuit Syst.-1: Fundam. Theory Appl. **41**(5), 46–49 (1994)

Author Index

© Springer Nature Singapore Pte Ltd. 2019
J. K. Mandal et al. (eds.), *Advanced Computing and Communication Technologies*,
Advances in Intelligent Systems and Computing 702,
https://doi.org/10.1007/978-981-13-0680-8

Printed in the United States
By Bookmasters